On the Eastern Front 1914

Meine Kriegserinnerungen

by Werner N. Riess

edited by Warren C. Riess

1797 House

copyright © 2020 By Warren C. Riess

All rights reserved. No part of this publication may be reproduced, stored in a retrieval system, or transmitted in any form or by any means—for example, electronic, photocopying, or recording without prior written permission of 1797 House. The only exception is brief quotations in published reviews.

Published by 1797 House
Bristol, Maine USA

Library of Congress Cataloging-in-Publication Data

Memoir Riess, Werner N. /Warren C. Riess, editor
On The Eastern Front 1914: Meine Kriegserinnerungen
Includes memoir and annotation in English, original memoir in German, illustrations, maps, references, and glossary.

ISBN 978-0-9713438-1-8 paperback
978-0-9713438-4-9 hardcover
978-0-9713438-3-2 ebook
978-0-9713438-2-5 Kindle

1. First World War, WWI, Eastern Front, German field artillery, 1914, memoir, kriegserinnerungen,

Library of Congress Control Number: 2020902757

Translated by Christa Mayer-Bohne
Covers and layout designed by Kathleen Galligan
Maps by Warren Riess

Covers: the leather cover of *Meine Kriegserinnerungen* (My War Memoir), including an embossed Iron Cross, 2nd Class.

Back cover: a flower that Werner Riess sent to his wife for her birthday, October 1914. She pressed it in her diary.

For the millions of people who suffered in and from the First World War

Contents

Preface and Acknowledgments v

Introduction vii
- A Brief Background on the First World War
- Editor's Note and Further Reading

Military Terms xv
- Glossary
- German Field Artillery Units
- Order of Battle

1. My Life as a Gentleman Was Over
   28 June–18 August 1914 1

2. First Blood: Gumbinnen and Tannenberg
   19 August–1 September 1914 27

3. Battle of the Masurian Lakes and into Russia
   2–26 September 1914 51

4. Feigning to Be a Solid Front
   27 September–10 October 1914 73

5. South, Along the Russian Border
   11 October–1 November 1914 95

6. Into Poland with the Ninth Army
   1–30 November 1914 123

7. Praying for a Silent Night
   1 December 1914–1 January 1915 147

Epilogue 171

Original German Text of Werner Riess's Memoir 175
of Werner & Gertrud's Poem 263

Preface and Acknowledgments

In the spring of 1915, Sgt. Werner Riess was recovering from surgery at a military hospital in Wiesbaden, Germany. Referring to his battlefield diary and letters that he had sent home to his wife, he wrote this memoir of his time in the German field artillery, fighting on the Eastern Front against the Russians in the First World War. The memoir starts on 28 June 1914, the day that Archduke Ferdinand and his wife were assassinated, and continues through 1 January 1915, when Riess returned to Berlin, suffering from chronic gallbladder disease.

Werner Riess related what he did, what he saw, and what he felt, sometimes using sarcasm to emphasize perceived irony. Because at the time he wrote the memoir he still did not know what had transpired out of sight around him, I have added annotation text to each chapter and included some pertinent material from his wife's diary. I hope these additions will help modern readers understand the background and "larger picture" of what Riess related.

When he finished writing his war memoir, Riess had two copies typed and leather bound in April 1915. The two bound copies have remained in the family ever since. They passed from his wife, Gertrud, to his only son, Herbert, and then to Herbert's children—my four sisters and myself. We have retained both copies. The original handwritten manuscript has not been found. My wife Kathleen then urged me to translate, edit, and publish it. The entire memoir is presented here, translated into English by Christa Meyer-Bohne, with my few changes to modern American usage to help the non-military reader. A scanned version of one of the original 1915 typed copies is included for reference.

We are all indebted to Werner Riess for writing this memoir and to Gertrud, his wife, for keeping Werner's memoir and her diaries safe. As noted, she passed these on to their son, Herbert Riess, who kept them safe through the decades since and passed them on to his children before he died. An American whose native tongue is German, Christa Mayer-Bohne not only translated Werner and Gertrud's writings but also researched early twentieth-century German military terms.

Many people helped my annotation by providing information about older German customs, early twentieth-century Berlin, and First World War military practices and equipment, especially Warrant Officer Ralph Lovett, US Army, retired; Dr. Stephen Miller, and Dr. David Patton. Master farrier Fred Bowers tutored me about the horses' needs.

Within Germany, Michael and Traudel Meyer-Reinhard were of great help with sources and translations. Tessa Norton and Kathleen Riess helped me with research in the German archives. Staff members of the Hover Library, Stanford University, and Randy Lackovic at the Folger Library, University of Maine, made primary and secondary sources easily accessible. Warrant Officer Ralph Lovett and Dr. David Reiss helped me understand the stresses Werner Riess and his comrades experienced.

Early readers helped me decide on the eventual structure of this work, including Dr. James Bradford, Revell Carr, Dr. George Daughan, Robbie Downs, Barbara Muller, Dr. Clint Schaum, and especially Roger King.

For editing my contributions to this manuscript, I thank so much Tessa Norton, Kathleen Riess, Patricia Tennison, plus professional editors Sally Antrobus and especially Marianne Stone.

Warren C. Riess, Ph.D.
Research Associate Professor, Emeritus
Department of History, University of Maine
Grandson of Werner N. Riess

Introduction

In July 1914 Werner Riess was twenty-six years old, married with no children, enjoying the life of a young urban gentleman. He was a department store manager in Berlin, but like most German men, he had served in the army and then in the active (ready) reserves.

This is the only photograph found of Werner Riess (center), with his father Guido and younger brother Kurt, c. 1902.

When the war began Riess was mobilized and assigned to a new field artillery battery, mostly supporting the infantry and sometimes attached to cavalry units. The German field artillery moved rapidly to reinforce with their firepower other units that were under attack or that were advancing on the enemy.

Riess served on Germany's Eastern Front, fighting Russians, from the first days of the war until the last day of 1914, when he became so sick that he was sent back to Berlin. Having served in a very mobile fighting unit, he left us with a personal view of much of the Eastern Front and many of the major battles in 1914. He never tried to publish his memoir. It is probable that he wrote it only to record for his family his involvement in what was then the largest war in human history.

Riess related the war as he saw and felt it, shortly after he experienced it. He wrote about the people and fighting immediately around him. He didn't give an overview of the battles, which sometimes covered hundreds of square kilometers, and he never described the whole Eastern Front. Not being a general, he did not know the perspective of the commanders, and he never knew the Russians' view of the Eastern Front. And so he has left us a seemingly honest personal memoir, not edited by the filters and analyses of historical perspective.

In each chapter, after his account, I have added clarification and annotation to present the background, details, and wider picture. I have been consciously German-centric to help readers understand Riess's point of view. Except for some overviews, I have discussed only that part of the war that directly related to Riess's experiences. I left out much about the rest of the war that one can learn from the many good books on the subject. However, little has been written about the Eastern Front of the First World War. Winston Churchill called it *The Unknown War*. Possibly because there was no victor there, few have written about it. What they did write was mostly about the generals and large unit movements. Here, a soldier on the front lines offers us a personal, close view of the constantly changing fighting between Germans and Russians during the first five months of the war, written while it was still fresh in his mind.

Reading Werner Riess's words, and hearing the young, patriotic voice of his memoir, I hope it conveys some of what millions of individuals, from many countries on both sides, experienced a century ago. They were in the most devastating war the world had known that, as Riess wrote, “was to bring unspeakable misery” to so many people.

A Brief Background on the First World War

Europe in the early twentieth century was a mix of old and new countries with intermingled cultures and various types of monarchies and republics. Their governments were the temporary results of centuries of intrigue and revolutions, with borders repeatedly redrawn by inheritance, conquest, and purchase. Most nations included land and people that had once, or more than once, belonged to neighboring countries. Beneath the surface of diplomacy and military power lay a web of intrigue with spies collecting information and a complex set of secret agreements that no one player knew. In the early twentieth century it was clear to everyone that Europe was a tinderbox that only needed a spark, and every country added fuel to the mix by increasing military preparations.

Germany, then a young empire, was formed by many of the German-speaking kingdoms in 1871 and ruled by Kaiser (Emperor) Wilhelm II of Prussia. The German military was a formidable power in northern Europe that still consisted of distinct units from each principality. By far, the Prussian Army was the largest contingent of the German Army. Though Austria-Hungary, to Germany's south, had lost the leadership of the German-speaking people to Prussia in 1866, the Austro-Hungarian Empire and Germany were allies in 1914. Each would come to the other's aid if they were at war. Italy by treaty would come to the aid of either, but only if that country were attacked first. Together Germany, Austria-Hungary, and Italy formed the central Triple Alliance.

Serbia, politically overshadowed by Austria-Hungary, held a treaty with Russia in which the Russians would militarily support them if Austria-Hungary invaded. Both France and Russia (which occupied much of Poland at the time) saw the Germans as a direct threat to themselves. By their treaties, if either were invaded the other would attack Germany; hence France and Russia threatened Germany with an eastern and a western front if the Kaiser started a war.

While most cultures are certain that they are the center of humanity, many Germans had also accepted a century of philosophy by Fichte, Hegel, Nietzsche, Treitschke, and others that painted the German people as especially superior. Prussian militarism, stressing the importance of possible aggressive military action in political decisions, had become an accepted part of the German culture that the French and Russians emulated with some success.

Most of the European military, with the support of many civilians, believed in the need for aggressive action whenever war would come. Many took to heart Prussian General Carl von Clausewitz's writings, especially the 1832 *On War*, wherein he asserted a philosophy about war in a modern world. He argued that a strong military was a necessity for any country that hoped to survive and flourish. Military and political leaders throughout Europe analyzed past campaigns and planned for the next using Clausewitz's principles of war, such as remaining alert and flexible because war is unpredictable, the importance of military morale, support by the total population, and the benefits of professional armies.

Through the late 1800s and early 1900s the European countries, including Russia, increased the size and quality of their armies. Tens of thousands of men expanded to hundreds of thousands of better-trained soldiers. Hundreds of older cannon were replaced by thousands of modern, quick-firing guns with deadlier ammunition. Hundreds of the newly invented machine guns and millions of cartridges were in their armories. Strong new underground fortresses, armed with modern artillery and machine guns and surrounded by barbed wire, defended the borders and larger cities.

France and Russia each had about the same number of men in their active armies as Germany. While French and German forces were considered second to none, Russia's military units were declared good but not on par with those of the western powers. However, the Russians, with their vast empire, could call up millions of reservists within a month. The Austrian forces were very good, but Germany realized that if the war came, Austria-Hungary would be tied up fighting on their own front against Russia and Serbia. Both sides thought Italy, because of its own interests, probably would not be a key player in any war.

The United Kingdom of Great Britain and Ireland, led by the English, was an undealt wild card for European war planners. The English were sometimes allies of the Germans, but they professed to be neutral in any continental war, except to support neutral Belgium if it were attacked. Belgium was too close to England's southern coast to allow an enemy to hold it. The English had a long history of keeping the European continental powers from invading their islands by siding with the underdog on the continent. England's small army and mighty navy might be important players if they became involved.

On 28 June 1914, a Serbian nationalist who wanted Serbia to be free of Austrian control assassinated Archduke Ferdinand, heir to the Austrian Empire, and his wife. All of Europe prepared for an intense, short war. Riess's memoir starts with that day.

Editor's Note and Further Reading

The chapters that follow present Werner N. Riess's account translated into English and rendered in Courier font, as was his original memoir, while my explanatory text is in an alternate font. A scan of the entire original memoir, in German, appears at the end of the book.

Many of the place names have changed from those Werner Riess used in 1914. In the text and maps I have used the same place names as he did, with his spellings, though even in 1914 people of different nationalities used various names and spellings for the same places. In parentheses I have included some of the modern names. I suggest keeping the map pages bookmarked, otherwise his constant traveling may not make sense. Similarly, I have not changed Riess's spellings of people's names.

Sometimes in his memoir, especially in the last two chapters, Riess mentioned quartering on an estate, sometimes referring to the manor or castle only by the estate's name, as in "That evening we took quarters at Wieslawice, a beautiful castle," in chapter 6. Today, after a century of cultural and political changes, most of these names refer to the same place, but it has changed from an estate to a village or town.

My annotation includes few citations because almost everything there is a compilation of information from well-known secondary sources about the war. The one hard to find book is the official history of the 36th Field Artillery Regiment, from which I borrowed many details for my text and maps: *Das Reserve-Feldartillerie-Regiment Nr. 36 in Weltkrieg.*[1] Most details, unless otherwise noted, come from Riess's memoir or from this regimental history.

[1] *Das Reserve-Feldartillerie-Regiment Nr. 36 in Weltkrieg* (Berlin, 1929).

For further reading in English, I suggest:

Prit Buttar's *Collision of Empires: The War on the Eastern Front in 1914* (Oxford, UK: Osprey Publishing, 2014) is an overview of the part of the war in which Werner Riess served, mostly from the view of the Austrian, German, and Russian generals.

Eric Brose's *The Kaiser's Army, The Politics of Military Technology in Germany during the Machine Age* (Oxford: Oxford University Press, 2001) presents a good analysis of the German military's path to the First World War.

Winston Churchill's *The Unknown War: The Eastern Front* (New York: Charles Scribner's Sons, 1931) is a good overview from a British perspective.

Paul Lintier's *My Seventy-Five: The Journal of a French Gunner August–September 1914* (Solihull, UK: Helion and Company, 2012) is a fine, sensitive memoir by a French soldier in a similar field artillery unit on the Western Front.

Nicholas N. Golovine's *The Russian Campaign of 1914: The Beginning of the War and Operations in East Prussia* (The Naval & Military Press, 1933) presents a Russian overview and excellent maps of the early Eastern Front.

Daniel Showalter's *Tannenberg: Clash of Empires* (Hamden, CT: Archon Books, 1991) gives an excellent presentation of the Eastern Front in August 1914.

Alexander Solzhenitsyn's *August 1914* (New York: Farrar, Straus, and Giroux, 1972), a novel, offers a personal view from the Russian side.

Barbara Tuchman's *The Guns of August* (New York: Random House, 1962) is still the classic for understanding the First World War.

Military Terms

From Christa Mayer-Bohne's translation I have made a few changes into modern American English to avoid possible misunderstandings. For example, I use *shells* for exploding Russian artillery projectiles, whereas Riess used the older term *grenades*—otherwise a modern reader might think the enemy was somehow sending hand grenades over kilometers of terrain.

Glossary

battery: In this memoir, a unit consisting of four 77 mm guns (Krupp M96na's) and approximately 100 men, 100 horses, the necessary supplies, wagons, and equipment.

bivouac: To sleep or rest without tents. The troops would try to find shelter under wagons, caissons, trees, or blankets as best they could.

caisson: A 2-wheeled cart hitched to a limber that carried ammunition, men, tools, and supplies. An ammunition caisson was positioned close to each gun.

cart: A two-wheeled (single axle) vehicle.

forced march: A long-distance march performed as fast as possible.

gun: In this book refers to a piece of artillery, rather than a soldier's rifle or pistol.

howitzer: A short-barreled cannon that shot its projectile in a high trajectory, to shoot over obstacles or into trenches, etc. In World War I the German howitzers were mostly 105 mm, 150 mm, and 211 mm guns.

hussars: Light cavalry used mostly for scouting and quick raids. The primary weapon of the hussars that Riess supported was a 3 m (10 ft) steel lance, but they also carried a sabre, pistol, and carbine.

jägers: Elite light infantry, roughly equivalent to American rangers, trained for scouting and raids into enemy territory. *Jäger,* sometimes *jaeger,* is German for *hunter.*

kilometer: 0.6 mile; 10 kilometers equal approximately 6 miles.

limber: The cart to which the horses were attached and that had a hitch for other carts or guns.

NCO: A non-commissioned officer, meaning a corporal or sergeant. Riess began the war with the rank of *unteroffizier*, which was roughly equivalent to an American or British corporal, in charge of ten to twenty men. I use *corporal.* After two months he was promoted to *vise-wachtmeister*, and then *wachtmeister*, which are usually translated to *cavalry sergeant*; the rank of *wachtmeister* was used by all German mounted units, including their field artillery.

regular army: Active, full-time soldiers and army units.

reservists: All able men in Germany were required to spend at least two years in training and active duty in the regular army or navy, then many years of reserve service. Cavalry and field artillery soldiers, like Riess, were required to spend three years in active duty. In peacetime, reservists were required to report routinely for refresher training.

squad: One of a field artillery battery's guns and its crew.

troop or platoon: Two of a field artillery battery's guns and their crew.

vehicles: All pulled by six-horse teams in the German field artillery.

wagon: A four-wheeled (double axle) vehicle.

German Field Artillery Units

Artillery in 1914 consisted of foot (heavy) and field artillery. In most armies at the time, heavy artillery shot a projectile more than 100 mm (4 in) in diameter for many miles. They were a mix of flat trajectory "cannon" and high trajectory heavy howitzers; fuller descriptions of "field artillery" and "howitzer" follow.

Field artillery units in all the European armies at that time mostly used a small, low-trajectory cannon to fire directly at the enemy at close range or light howitzers to fire over obstacles.

The low-trajectory guns had a steel shield to protect the gunners from enemy bullets and were drawn by six horses. They were in the battlefield, usually just behind the front line of infantry to fire at machine gun nests, other field artillery, and concentrations of infantry and cavalry. Some countries used the term *horse artillery* for similar units equipped and specially trained to accompany cavalry. However, Germany equipped and trained all field artillery units to support infantry or cavalry and called them all *field artillery*. Because German field artillery units were mounted—that is, horse-mobile—the field artillery soldiers' ranks were the same as those of the cavalry.

The Krupp M96na field cannon was the most common rifled cannon for Germany's field artillery in 1914. It was Germany's first field cannon to have very little recoil, allowing the crew to stay near it when firing and quickly reload, aim, and fire again. It fired a 77 mm (3 in) diameter, 6.8 kg (15 lb) high-explosive or shrapnel shell in a low trajectory directly at the enemy, to a maximum of just over 8 km (5 mi).

The 36th Field Artillery Regiment, in which Riess served, was armed only with the 77 mm M96na. They mostly supported the division's infantry. Possibly because of extra training, when the cavalry required artillery support, the regiment's 3rd or 6th Battery accompanied them. Riess was in 6th Battery.

Some German field artillery units also used 105 mm "light" howitzers that delivered indirect, heavier projectiles than the direct-firing 77 mm M96na's.

Order of Battle

Military units change their structure as resources and missions demand. Riess was in 6th Battery of 2nd Battalion, 36th Reserve Field Artillery Regiment, 36th Reserve Division (Infantry), 1st Reserve Corps, of the German Eighth Army.

The 1st Reserve Corps was a new entity that consisted of newly created units of ready reservists, mostly commanded by regular army officers. After three months of fighting in the Eighth Army, in November 1914, 1st Reserve Corps was reassigned to the German Ninth Army.

On the opposite page I list the general structure of the German Eighth Army as the fighting started, leading down to 6th Battery, the unit that included Werner Riess. The Eighth Army was manned almost entirely by Prussians, therefore it also was known as the Prussian Eighth Army. The numbers of men are approximations, as those numbers changed daily. Riess was in the units indicated in bold type.

6th Battery had four Krupp M96na field cannon, approximately 100 men and officers, 100 horses, ammunition/baggage/supply carts and wagons, and other supporting equipment. Though 100 men might seem like a high number for a battery of four guns, a battery was designed to be a complete fighting unit that could operate entirely on its own for at least a day. Each battery was led by a captain and included several lesser officers, sergeants, corporals, four gun crews, horse team drivers, ammunition and supply squads, cooks, a veterinarian, farriers, stretcher bearers, and a repair team for the guns, wagons, and harnesses.

Riess began the war as a corporal. When 6th Battery was formed he was placed in command of 6th Battery's supply team, including approximately 30 men, 30 horses, and six wagons. Private Sell was assigned to be his orderly.

German Eighth Army—aka Prussian Eighth Army

(200,000 men)

1st Corps

17th Corps

20th Corps

1st Reserve Corps (38,000 men)

1st Reserve Division

3rd Reserve Division

Höherer Landwehr-Kommando No. 1

1st Cavalry Division

36th Reserve Division (15,000 men)

69th Reserve Brigade

70th Reserve Brigade

54th Infantry Brigade

1st Reserve Hussars

36th Reserve Field Artillery Regiment (24 guns)

most of the 2,000 men in the regiment came from Berlin

1st Battalion (1st, 2nd, and 3rd Batteries)

2nd Battalion (official list, August 1914: 658 men, 641 horses)

4th Battery

5th Battery

6th Battery (100 men)

Map 1. Eastern Germany, (Prussia) showing the area Riess transited, 4–18 August 1914.

Königsberg
Insterburg
Gumbinnen
Nordenburg
Angerburg
Marienwerder
Allenstein
GERMANY
RUSSIA
Tannenberg
Thorn
Vistula River
POLAND
Warsaw
Lodz
N
0 50 100 km
0 30 60 miles

-1-

Am 28. Juni 1914, es war ein herrlicher Sommertag - in Hamburg wurde das Deutsche Derby gelaufen, bei dem ich ja seit Jahren niemals fehlen durfte - wurde der Grundstein zu dem grossen, grossen Weltkriege gelegt, der über uns alle, über Freund, wie Feind, so unsagbares Unglück bringen sollte und der, während ich heute in der Mitte des April 1915 diese Aufzeichnungen machen will, noch in vollem Gange ist, der an Stärke gewiss noch nicht abgenommen hat, sondern eher noch zunehmen wird, denn Entscheidungen von grosser Tragweite, oder wenigstens von entscheidender für die eine oder andere Partei, sind noch nicht gefallen. Aber unsere Lage ist eine gute und der endgiltige Erfolg wird uns wohl ziemlich sicher sein.

Damals in den August-Tagen, als wir immer von neuen Feinden hörten, immer wieder hörten, dass neue Nationen, von denen wir zum grossen Teil angenommen hatten, dass sie in einem Weltkriege unsere Partei ergreifen würden, sich auf die Seite unserer Gegner gestellt hatten, haben wir uns draussen gefragt: Wie kann Deutschland allein alle die Völker, die sich ihm entgegenstellen, niederringen ? - Aber bis zum heutigen Ta-

Scan of the first page of Werner Riess's memoir, as typed in April 1915. A scan of the complete original memoir comprises the last third of this book.

CHAPTER 1

My Life as a Gentleman Was Over

28 June–18 August 1914

On 28 June 1914, a beautiful summer day, while the German Derby in Hamburg was taking place, an event I hadn't missed in many years, the foundation was laid for the great, great war that was to bring unspeakable misery to us all, friends as well as enemies. While I am writing down these notes today, in the middle of April 1915, the war is still going on, has not lost any of its strength and probably will intensify, as no major decisions or victories have yet been claimed by one or other of the parties involved. However, our position is a good one and a final success on our part is quite certain.

During these August days, we continuously heard that we had new enemies. We heard that nations we had earlier assumed would be on our side in a world war had changed to the side of our enemies and we asked ourselves openly, "How can Germany alone overcome all these nations that oppose it?" Up to this day, however, we Germans have demonstrated that we are able to do it, that we can withstand all our enemies because of our courage and bravery, our endurance of greatest hardship, our patriotism, and the unconditional trust we all have in our Supreme Commanders.

Even on that memorable day of 28 June, when suddenly news sprang up all over the racetrack that the Austrian crown prince and his wife had become victims of an abominable murder, there was a general consensus that we had to expect far-reaching consequences. Nobody would have thought then that the news would be cause to the most horrible conflagration that history ever had to report in this world.

By the beginning of August we knew that we were facing the start of a great world war that we had dreaded and avoided for many years, and that we were unable to comprehend. But now it was here anyway. The German Emperor, who had always wanted to preserve peace for his country, who had shown his love for peace throughout and who had wished to be remembered above all as the peace emperor in German history, was now to become the emperor to lead Germany into the mightiest war.

What did it feel like here in Berlin those days shortly before and after the declaration of war? There was great enthusiasm everywhere so that we felt reassured that victory was a sure thing! On the other hand, we all felt great concern about all we would have to leave behind. The days before my departure were all but pleasant. Leaving was extremely difficult. Much more difficult than I have ever wanted to admit to myself. We knew little of anything at that point, didn't know where we were going, didn't know if we would meet people out there who would be sympathetic, who would take an interest. Would we see again all those so dear to our heart?

I had to leave home on the third day of general mobilization, 4 August 1914. It was hard, terribly, terribly hard. According to my orders I was to show up at a certain schoolyard at 5 o'clock in the afternoon. When the first "comrade" I ran into (he wasn't wearing a collar and had slippers on his feet) jovially patted my shoulder and asked: "Tell me, where exactly are we supposed to gather?" I realized right away that my life as a gentleman was over for quite a while. I have always been one who is able to adapt well in different situations and so we two set off together toward that schoolyard from which our convoy was to leave (I never saw him again any time later and never knew who he was and what became of him).

4 August: Berlin to Marienwerder! En route from 8 o'clock in the evening till 9 o'clock the next evening (normally I think it takes a regular train about 8 hours). There were stops in even the tiniest villages and each time refreshments were handed to all of us heading to the field. We were tempted with good things to eat offered by those beautiful and not so beautiful ladies of West Prussia.

In my opinion it was almost too much of a good thing to keep our stomachs from aching and keep us in good health.

I rode in a second-class coach car along with the convoy leader, a captain of the medical corps, a pharmacist, and two vice cavalry sergeants. What happened to the officers I do not know, but both vice cavalry sergeants ended up in my unit. One of them, Vice Cavalry Sergeant Schröter was still there when I left.

The other, Vice Cavalry Sergeant Masch, who by profession was an opera singer in Stettin, lost both legs and one arm after his battery commander, Captain Hamann, feeling very "kindly" toward him had him transferred to the 25th Corps. I mention his fate so early on because he was one I went out into the war with. How thankful I am to the Lord that I can sit here with healthy limbs, while he, who probably would not have lived in the best of circumstances with his young war bride, now, if he still lives, will have to endure a very hard existence, harder than anything imaginable. But surely he is only one of millions who are enduring the same fate, and that is what is so depressing every morning when we get up and every night when we go to bed: we continue to lead our old quiet life—and out there the world is bleeding!

But now let me return to my account! Well then, we arrived in Marienwerder on 5 August and there I was in luck for the first time in this campaign, and luck did, thank God, not leave me. The major who was leading our unit was my old Captain Kujath, who had very much liked me in the days of my military service. He recognized me immediately and it was very pleasant to know that there was a human being who had previously shown some interest in me and who knew what family I came from.

We were put into batteries that same evening and then all of us reserve NCOs [corporals and sergeant] were allowed to go to a hotel. On entering the hotel room I thought to myself, "Oh my God, so that's the start of the war!" The crudest of crude places, but it would have felt like the bedroom in a prince's palace if I could have had it later. From 4 August until 31 December, the length of time I was in the campaign, there were perhaps fewer than ten occasions when I was able to lay my weary limbs in a real bed, sleep in a "comfortably equipped" room, and in the morning wash myself in a proper washbasin and have breakfast at a table.

In the afternoon of the 5th, after we had been issued clothing in Marienwerder, we went to Gross-Krebs. I had already become friendly with my Cavalry Sergeant Martin and he took me to his quarters where there was bedding for me in the attic among apples, pears, mice, and rats. It was really comfortable though, because at that point one could still get undressed! Food and board were excellent while we were at Gross-Krebs and we surely did not endure any bad times there.

On the 10th we were armed and equipped; and with our weapons went from Gross-Krebs to Marienwerder and were entrained to go to Nordenburg. We arrived in Nordenburg in the evening of the 11th and after marching for about 2 hours we arrived at our first encampment in Wilhelmssorge. Next morning we went from Wilhelmssorge to Bidaschken where a few thousand soldiers were stationed. Bidaschken was a small village of only 30 to 40 buildings and housing was accordingly poor.

People there didn't have anything to eat any more and the war had really started for them. Luckily, we didn't stay there for very long and on the 14th we were at Paulswalde.

I had good days there with the Family Stadthaus who I think of in friendly terms and who treated me very kindly. Old Mrs. Stadthaus would say, "We want to give up all we have as long as we can be certain that the Russians will never show up here." And we, who felt so sure of having the upper hand, replied that it would be impossible for the Russians to beat us, we were too mightily strong. But only a few days later, on the 20th, the Russians held Angerburg and therefore Paulswalde. And when I asked refugees from Paulswalde later if they knew what had happened to my dear old hosts, I was told that they too had to flee just like everyone from that area. I only hope that the Russian hordes did not completely destroy their beautiful property.

We worked hard at Paulswalde, did gun drills, aiming exercises and so forth. Earlier on I had been given command of our combat baggage and therefore had very little to do with these drills. Instead I spent a great deal of time in Angerburg to make acquisitions for our battery and therefore enjoyed a few nice days. Upon returning to my quarters I was spoiled too. I had to myself two large rooms while four infantry officers shared an attic. They did not, however, know anything about my two rooms, otherwise I would have been out of there in no time.

An alarm was given on the 19th in the morning and we moved to entrench a few kilometers away from Paulswalde in the proximity of Sobiechen. We were told that this is where we were going to be for the next few days and therefore, toward evening, I went back to Angerburg to make more acquisitions.

****

For twenty-six-year-old Reserve Corporal Werner Riess, August 1914 was a time of dread. Enemies faced Germany on the eastern and western borders with huge armies. He was leaving his loving wife of four years and his genteel middle-class life in Berlin as he prepared to defend his family and his country. Although he was patriotic and seemingly sure of the German army's abilities, his earlier three years in the army had taught him something of the meaning and uncertainty of war, and he wondered what it would bring for him.

We know little of Riess's early years, except that he came from a mixed Lutheran and Jewish family in Stettin (now Szczecin, Poland), on the Baltic coast northeast of Berlin. Stettin was originally a Polish town that passed between the Poles, Swedes, and Prussians several times through the millennia. In the late nineteenth century its people were a mix of Germans, Poles, and Scots—mostly Lutherans, many Roman Catholics, and some Jews. By his twenties, if not from birth, he was a practicing Lutheran. His family had a small furniture factory and store in Stettin and they owned a tree farm to supply wood for the furniture.

At nineteen years old, Riess began his required three years of service in the German Army. Possibly because of keen eyesight, a mind for mathematics, and riding abilities, he was assigned to the field artillery. They were trained to travel, maneuver, and set up their small cannon rapidly to give close support to infantry or cavalry units with accurate, direct fire. After his three years, he was assigned to a ready reserve field artillery regiment that regularly met for training.

In August 1910, shortly after his active duty, he was living on Poststrasse, Mitte (the central district of Berlin), and married eighteen-year-old Gertrud Grumach, a neighbor. Her father and uncle owned Gebrüder Grumach (Grumach Brothers), a five-story department store just a short walk away. Gertrud's family was part Jewish German and part Scottish, but she, like Riess, was a practicing Lutheran. We do not know if theirs was an arranged marriage; or a boy-meets-girl story, since they lived two houses apart on the same street; or if Riess married the boss's daughter. Shortly before or after their marriage, Riess became a manager in Gebrüder Grumach. The newlyweds moved to an upscale neighborhood in Schöneberg, a southwest borough of Berlin. With a good income, and possibly helped by their families, they were able to rent a large apartment and hire a cook and a maid.

As the capital of the German Empire, Berlin was the official home of the kaiser and the upper echelon of the military, political bodies, and civil service. In the days leading up to and after the declaration of war on 1 August, Riess and his wife would have heard many false rumors of peace and war that coursed through Berlin—sometimes reassuring and sometimes unsettling. Berliners in the early twentieth century were a mix of conservatives, liberals, nationalists, and socialists. The city had grown quickly in the last three decades and felt the strain of a rapidly expanding modern city with new industry, service businesses, and wholesale and retail businesses.

Most of Berlin's buildings were new and efficient. Some called it Germany's Chicago. Some called it spiritless and inhuman. As in many societies, rapid changes brought internal tensions, political polarity, and hostility. By early 1914, stresses in Germany were separating Germans into the political extreme right and left, especially in Berlin, and the socialists were gaining political power.

It is not evident from either their written documents or family lore how Gertrud and Werner felt about the political atmosphere. By 1914, just before the war started, Riess was a partner in Gebrüder Grumach.

Like many Europeans, the young couple thought that the late June assassination of Archduke Ferdinand and his wife would bring problems for Europe, and perhaps a quick, bloody war. However, as Riess stated in his memoir, the sequence of events that followed was unexpected for almost everyone, even those in the diplomatic and military communities. That July, millions of people in Europe, and many more around the globe, watched as Austria-Hungary and Serbia mobilized for war. Many countries readied for an intense, three-to-six-month European conflict. Every European nation put its regular and reserve military on alert.

Knowing he would be leaving home if the army's reserves mobilized, Riess and his wife could only watch, wait, and worry during that summer as Europe tumbled into war. We might have seen him cutting short his straight brown hair, exercising, and double-checking that his horses and personal items were in perfect condition. During the summer he suffered a gallbladder attack and was sent to Karlsbad, the German name for Karlovy Vary, Czech Republic, to "take the cure." The regimen consisted of early rising to a large glass of mineral water, hot baths, a strict diet, and hours of rest punctuated with glasses of mineral water. His wife wrote in her diary that when he appeared for his military medical exam that summer he did not mention his ailment.

During the next weeks of the summer of 1914, in defense of Serbia, Russia declared war on Austria-Hungary. Then in defense of Austria-Hungary, Germany declared war on Russia and France. Immediately it became clear that almost every European country, large and small, would soon be engaged in war on one side or the other. In the first days of August every continental army quickly mobilized reserves for war.

While European nations deployed their armies for war that summer, the kaiser and his advisors hoped that, among other things, the coming war would rally all German citizens around the kaiser as the symbol of the empire, and it did. By August the several political parties, left and right, all but stopped functioning, while the populace in every kingdom of the federation united as Germans to fight their old enemies, the Russians and French. Protestants, Jews, Catholics, nationalists, and socialists from the cities and countryside rapidly transformed into German soldiers. Hundreds of thousands of individuals reported for duty to protect their loved ones and the fatherland.

Most reservists reported for duty in their civilian clothes, as they were to be issued new wool field uniforms in a few days. Riess, being a wealthy middle-class gentleman, did the same but carried his own custom riding boots and personal decorative pickelhabe, or spiked helmet. The German helmets were made of black boiled leather. His would have sported a large silver-plated Prussian eagle in the front and, because he was in the field artillery, an artillery knob on top, rather than the more common spike.

From Werner's and Gertrud's diaries we know that he left home with many small gifts from her: writing paper, pen, a journal, some of his favorite cigars, and a photograph of her:

In her diary she wrote about the photograph, "On it I wrote holy words of great devotion and great love." She pasted a copy into her diary.

Gertrud probably packed for him changes of underwear and socks, handkerchiefs, and toiletries. He carried cash with him to pay for personal expenses. He had already delivered his two beloved horses, Intarsia and Fuchsrot, to the depot. As both names refer to an animal's color, Intarsia was probably multicolored, possibly shades of brown and chestnut, while Fuchsrot may have been a ginger or chestnut color.

We see in his words a love of family, patriotism, and then the excitement of being with old and new comrades in arms. As Riess gathered with the other reservists at the train station on 4 August they were accompanied by thousands of well-wishers and sightseers. Many carried flowers, a traditional gift given to warriors mobilizing for war. Some reservists carried their dress uniforms in their packs, thinking there would be a victory parade once they smashed the enemy; but which enemy? They did not know whether they would be fighting on the Eastern Front against "the Russian hordes" or on the Western Front against France's elite troops. This became clear, however, when the train left Berlin. They were heading east.

Riess wrote that he was in a second-class train car. In 1914, first- and second-class train cars were divided into compartments with a door leading to an interior hall and another to the outside. Seating was on two upholstered benches, with a luggage rack and horizontal mirror over each bench. He settled into his seat and made new acquaintances—one was an opera singer from his hometown of Stettin.

We can assume that during their long train ride, the reservists exchanged tales and information about their families and their previous army exploits, all the while smoking and consuming food and drink given to them by their loved ones left behind and well-wishers along the way. Aromas of fresh bread, sausage, cheeses, and beer would have mingled with those of tobacco smoke and newly treated leather. Dust and perspiration would have permeated the train as August 1914 was a very hot, dry month in Central Europe.

Most of these soldiers were new to war. Internally producing waves of adrenaline and testosterone, they must have felt uncertain, with swinging moods of bravado, apprehension, and being in the hands of God or fate. Without previous experience in battle, many would wonder "How will I react in combat?" and "What will my comrades think of me?"

There were things to discuss on the train that day, some hotly. Riess and his comrades, like soldiers throughout Europe, were probably arguing about the best use and power of that state-of-the-art weapon, the machine gun. These guns were heavy and slow to move, but they could lay down a terrific rain of firepower toward charging infantry and cavalry. Some men possibly argued that German field artillery could easily take out enemy machine guns because the Germans relied on accurate, low-trajectory guns. Others would have argued that the Russian long-range artillery might take out the German field artillery before it could do that. But as the train slowly made its way toward the Russian border their conversation must have shifted to the shocking news about "the English."

A few days before, Britain had warned everyone that if any country attacked through neutral Belgium, the British would fight to protect the small country. When Germany tried to sweep through Belgium to attack France, London gave them an ultimatum, which the kaiser waved off. He did not think Britain would really fight them because of something as insignificant as Belgium's neutrality. But at 8:30 p.m. on 4 August, thirty minutes after Riess's troop train headed east, the population of Berlin heard that the British had declared war on Germany. Mobs quickly formed in the streets, attacking people they thought might be British. British citizens in Germany were advised to wear American flags to pass for neutrals. When they heard this news along the way at one of their many stops, Riess and his comrades must have been enraged and disheartened to think that the British would suddenly turn against them for a deed that was so necessary to winning a war within which they were outnumbered by at least two to one.

As the reservists traveled into East Prussia they were proud to be members of the powerful German Eighth Army. However, their generals had little faith in the coming performance of reserve troops. Although they had trained and equipped the reservists well, military planners thought these men had grown soft and probably would not follow orders as well as currently active soldiers. The reservists also did not have the coherence of a team that had worked together for years or even weeks.

Constructing an army that would have to hold the Eastern Front against everything Russia could throw at them, the generals created new reservists-only divisions, rather than dilute the regular army divisions with questionable part-timers. Initially they would use the reserve divisions only in less important roles until the reservists could prove themselves.

At Marienwerder, Riess was placed in the newly formed 6th Battery of 2nd Battalion, 36th Reserve Field Artillery Regiment. The regiment was part of the 36th Reserve Division of the 50,000-man 1st Reserve Corps. Once outfitted in his new greenish-gray field uniform, Corporal Riess probably looked quite formidable with his black leather helmet exaggerating his sturdy 5-foot, 11-inch frame, and wearing his dull green field coat, leather pistol belt, canteen, and backpack.

The German army did not have enough equipment and men to bring the reserve units up to regular army numbers. Unlike the regular army field artillery, 6th Battery had four instead of six cannon, no panoramic sighting device to lay accurate fire at long range, and there were no kitchen wagons for the regiment to provide hot meals. Though pistols and carbines normally were issued to field artillery soldiers, none were available for the new reserve unit. They entered the war without any personal weapons. The division commander therefore had to assign a small unit of jägers (elite light infantry) to each battery to protect them.

6th Battery consisted of approximately 100 men: gun handlers, a stretcher-bearer, a captain, three lieutenants, three cavalry sergeants, four corporals, cooks, and men to repair the guns, wagons, and harnesses. The battery had its own combat ammunition and baggage wagons that could move quickly with them. The guns, supplies, and men were pulled or carried by approximately 100 horses; therefore the battery also included farriers with their tools and supplies.

At Gross-Krebs, a small village about 6 kilometers east of Marienwerder, members of the battery were issued their only weapons: four new 77 mm M96na cannon, considered a light gun. The M96na was a potent low-trajectory gun, weighing about a ton, mounted on a two-wheeled carriage, and pulled by six horses. The gun was similar to those of other modern countries. With it, a well-trained crew could accurately deliver a powerful exploding shell out to 5 kilometers every five seconds. The maximum range was over 8 kilometers.

The six horses could pull the 77 mm gun at a full gallop, though the battery usually marched at a walk or slow trot. A soldier rode on each of the left three (lead) horses. The driver, or teamster, was the soldier on the last horse, closest to the limber and gun carriage. All six horses were saddled in case the horses had to be shifted quickly or they needed to carry some of the wounded. Two gun handlers would sit on the limber, facing forward. Two more gun handlers could sit, facing the rear, on steel seats that folded out from the rear of the gun carriage. Other soldiers would be on the horse teams that drew the ammunition and baggage wagons and caissons. The officers and NCOs rode separately on their horses so that they could easily travel throughout the battery.

German field artillery battery practicing with their M96na's just before the First World War started.[2]

With steel shields on their mobile guns, the field artillery typically fought just behind the front line of infantry or cavalry. Other countries separated their field artillery, which supported their infantry, from their horse artillery, which supported the cavalry. The Germans had no designated horse artillery units. Instead, they equipped and trained their field artillery units for both duties. All their field artillery guns were drawn by six horses that could gallop while pulling the gun and could be easily ridden. The gun crews

---

[2] Köhler, Max, Der Aufstieg der Artillerie bis zum Grossen Kriege, 1938, Barbara Berlag Hugo Meiler, Munchin, p.137.

could set up carefully for accurate fire when they needed to fight with the infantry and could maneuver and fire quickly when attached to the cavalry. Setting up to fire involved:

- turning the six horses in the correct spot to have the gun facing the enemy;
- unlimbering—that is, unhitching—the gun from the rear of the limber;
- setting the recoil arms out, loading, and aiming; and
- meanwhile, the horses and limber would be taken 100–200 meters to the rear.

Artillery men typical referred to all these steps as unlimbering. Limbering was to reverse that process to be ready to move.

As part of mobilization, each of the horses needed to be inspected by the veterinarian and farriers, and 36th Regiment needed to hire civilian farriers to help. Also, most of the horses had to be trained for their military duties. Germany, like some of the other European countries, had a national system to supply the hundreds of thousands of horses needed in the event of war. The German Army bred, then raised and trained young horses for military duty. Some were bred for riding; some for towing field artillery guns, caissons, and light wagons; some to haul very heavy loads. Once they matured, the thousands of horses that were not needed by the regular army were lent out to private citizens, such as farmers, who used and took care of them. If war came, the horses were collected. The army also conscripted many privately owned horses and paid the owner an appraised sum. Some owners who were heading into the war, like Riess, took their own horses with them. Otherwise the army might conscript their horses, and then the owner might be issued an inferior mount at the front. All the horses, owned by the army or privately, had to be trained quickly to military duty under duress.

Near Sobiechen, while they readied their horses and waited for action, the field artillerymen practiced limbering, unlimbering, and firing their guns. The supply teams practiced rapidly acquiring, transporting, and dispensing ammunition, food, and water. The men accustomed themselves and their horses to the routine and extreme noise of the high velocity cannon and rifle fire. Without modern protection, the men's and horses' ears must have suffered terribly. In addition, they practiced signaling with hand signs and bugle calls that could be seen or heard over the din of battle.

With a few days of training, 6th Battery, along with the entire German Eighth Army, were prepared to face the Russians. Yet no one was sure just how the Russian hordes would deploy. The Germans knew that the Russian military was composed of many more regular troops than Germany could field. Yet 90 percent of Germany's forces were 500 kilometers to the west, facing the French.

The Russian army was heavy in cavalry, infantry, and artillery; but it had fewer supply units per division than the Germans. Frontal assaults by waves of cavalry might be used in the coming war, but most people felt the cavalry would be used to scout, or make swinging flank attacks, while the infantry and artillery hit the enemy's front. Russian artillery units were capable of long range fire but at slower rates of fire than those of the Germans. Their regular infantrymen were considered brave and stoic but poorly educated and poorly led.

The German soldiers facing east were trained to expect an attack from Russian heavy artillery, followed by massive frontal bayonet charges. They expected entire divisions of Cossack cavalry trying to sweep around and attack their flanks and rear. If, instead, the German generals ordered their troops to attack the Russians first, their plans were to hit the Russians with accurate heavy and light artillery. Then waves of German infantry would attack, while their cavalry would maneuver to strike the Russian flanks. These were their plans, and how they maneuvered when training, but no one knew what would happen.

By mid-August the French, German, Austrian, and Russian generals mostly were following their prewar strategies. Austria had invaded Serbia and was fighting both the Serbs and a Russian onslaught. In the west, as Riess mentioned in his memoir, Germany had invaded neutral Belgium, in order to swing around France's northern flank toward Paris. At the same time, while defending in the north, France massed its army and attacked Germany's southern flank through Alsace. Two Russian armies marched west to attack Germany's East Prussia.

On the Eastern Front only the 200,000-man German Eighth Army was sent to stop the Russian invasion until reinforcements could arrive. However, German reinforcements could be sent only after the western armies defeated the French. The Russians began invading East Prussia with their First Army to the north of the Masurian Lakes, and their Second Army to the south of the lakes. Their plan was for the Russian First Army to hit the main body of the German Eighth Army head on and keep them occupied. Then they would swing the Russian Second Army up from the south, behind the Germans, thus enveloping the 200,000 Germans with 400,000 Russians.

The German and Russian leaders had a pretty good idea of what their respective enemies had planned, but they understood all too well Carl von Clausewitz's fog of war—the inability to collect and analyze strategic and tactical information quickly, leading to uncertainty for commanders. The enemy might have leaked bogus plans to them, any of several variables might make the opposing generals change their minds, or poor weather or a lack of some resource might force a shift.

The German generals planned to repel the Russian First Army in the north when it attacked the German Eighth Army's defensive line, then swing south and attack the invading southern Russian Second Army. However, both battles had to be decisive victories, and the swing to the south had to be precisely timed and conducted. If it were not, the Russians would have a clear path into central Germany. Except for the German Eighth Army, there was no German force to stop the entire Russian military from invading all the way to Berlin.

While the German generals knew the Russian plans and tried to coordinate their various units as they gathered information from scouting cavalry and airplanes, line officers and soldiers knew nothing about it. They heard many rumors but knew only to follow their orders and always be ready to move, dig in, or fight. On 17 August at Stallupönen, forward units of the German Eighth Army fought with and forced back some invading units of the Russian First Army.

On 19 August the men of 6th Battery were near Angerburg, East Prussia, on the right (southern) flank of the German Eighth Army, entrenched and waiting for orders. Riess was in Angerburg with the combat baggage caissons and wagons, gathering last minute supplies from the depot for his battery. Only a few kilometers away, the Russian First Army of 200,000 soldiers were about to strike the German front line from the east, and the Russian Second Army of 200,000 soldiers were moving in from the south to attack them from the rear.

* * * *

A little over a year after the war started, Werner's pregnant wife Gertrud wrote to their unborn son about 1914. This is a translated excerpt:

Berlin, November 1st, 1915

It was in the month of August five years ago that your dear father and I celebrated our marriage. There were years of utter happiness, gaiety and blissfulness until August of last year arrived, this terrible August 1914. That is when the war began. It separated us for long, difficult months. When we had to separate it was for an indeterminate time and we had to think of the worst. A dreadful feeling overcomes me to this day when I think of those last hours before our separation into uncertainty.

On August the 1st a general mobilization had been proclaimed and on the 4th of August my dear husband had to depart too. What happened to him from that day on up to his return he has written about in great detail. Of me I can say little. It was a terrible time of never-ending anxiety and many ups and downs. Often, I didn't have news for long periods and this was especially hard. Bad foreboding would overcome me, the days were unsettling and filled with constant waiting for news. I didn't dare leave the house, afraid to miss the moment in which a sign from him would perhaps arrive. When news did arrive, sometimes just a few short lines, sometimes written a long time ago, then everything changed for a short time.

I have saved all letters from this time, even the shortest notes and my husband has done the same with his. These letters written in the mood of that time give the most accurate picture of the happenings and how we both felt.—However, there were many things I did not mention in these letters to my dear husband. Nothing about the sleepless nights, the hours of desperation and apprehension. Nothing is written about the illnesses I went through, the severe pain I found hard to manage given the poor condition of my nerves and my mind. It would have been unfair to burden him too with these details.

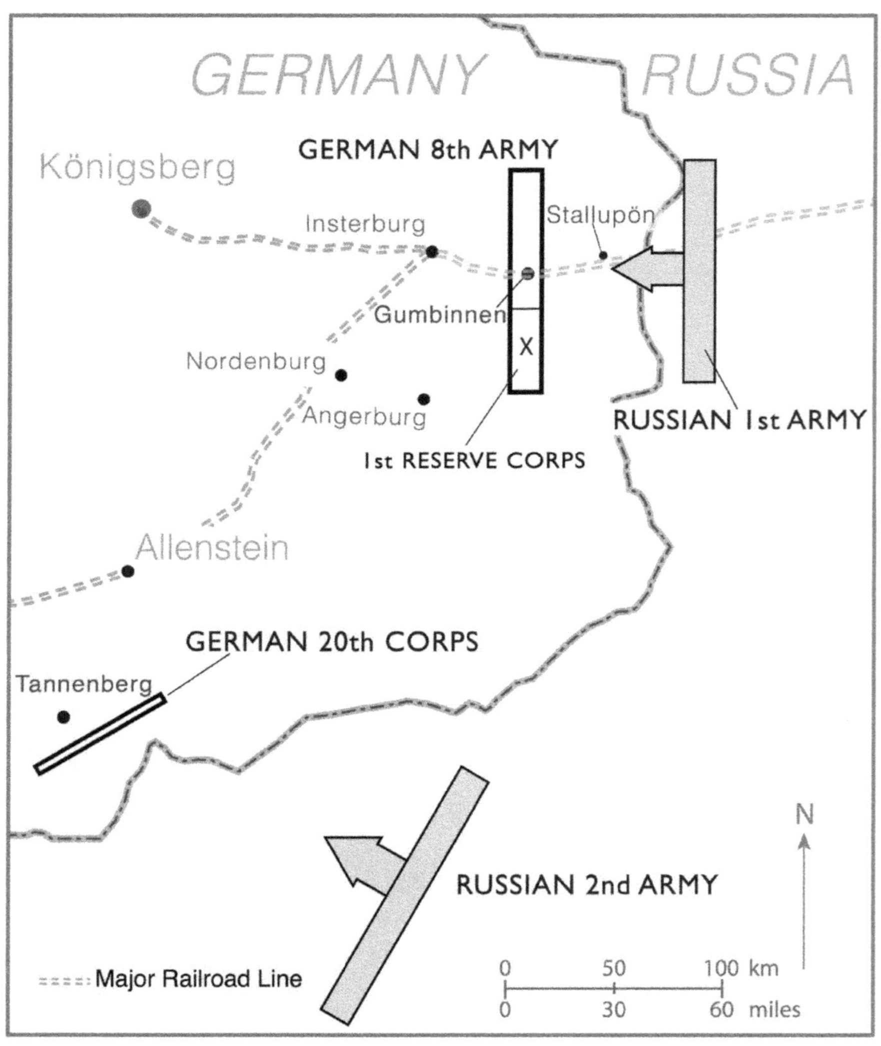

Map 2a. Approximate positions of the German and Russian armies, 16 August 1914. Riess was in the 1st Reserve Corps, northeast of Angerburg. One quarter of the German 8th Army, 20th Corps, was stationed near Tannenberg to hold the Russian 2nd Army.

CHAPTER 2

First Blood: Gumbinnen and Tannenberg

19 August–1 September 1914

19 August: When I was back [from Angerburg] within the vicinity of Sobiechen that night one of our buglers caught up with me and explained that the battery had had to leave in a hurry and that we should either try to hook up with our battery or with the large convoy. I succeeded in doing the latter and stayed with the large convoy as our corps was engaged in heavy fighting.

Wild rumors sprang up; we heard of great victories and half an hour later great losses were reported. As it turned out there was no great victory because on the 23rd we were taken as far back as Nordenburg again. The opinion of the leader of our army, von Prittwitz und Gaffron, was that we should move all the way back to the River Weichsel and leave all of beautiful East Prussia to our enemies. But thank God, our Emperor wanted something different.

I remember that it was announced on the 23rd that we were to get a new leader, a Mr. von Benekendorf. Who that Mr. von Benekendorf really was we found out a few days later after our great victory near Tannenberg became public. Very few people today are aware that our great Field Marshal von Hindenburg in former days carried the name von Benekendorf and that his full last name is von Benekendorf and von Hindenburg.

But now let me continue! Our convoy was at Nordenburg on the eve of the 22nd. Stories were told that there was nobody left in our battery and that there was only the captain and one cavalry sergeant left in the 5th

Battery. Was that really true? Not a word of it! Our battery hadn't lost a single man and the 5th lost one who had fallen off his horse. Nowhere are there more false rumors spread than in the war. We were very happy to see our comrades return unharmed.

On the 23rd we were supposed to have a day of rest at Nordenburg. But at 11 a.m. already an alarm was given and we had to encamp close to Nordenburg during really beautiful weather. On the 24th I noted in my diary: "Withdrawal, encampment." How far back we retreated that day I cannot say. I only know that we were taken forward again on the 25th because our corps was supposed to help the 20th Corps, which was involved in heavy fighting. By midday the news spread that the 20th Corps had won a handsome victory over the enemy. That was the first day of the mighty battle at Tannenberg, in which we had not participated actively until then but were going to be very much involved the following day.

On the 25th we also received news that Italy had declared war against France and Russia. Today, as I read in the newspapers about some big demonstrations Italian students are leading against German professors, I can but smile. But back on that day we shouted a mighty hurrah for Italy. Too bad!

Map 2b. Approximate positions of the German and Russian armies, 25 August 1914. Riess was northeast of Allenstein, in the 1st Reserve Corps, which was ready to assist north or south.

On the 26th our corps was also selected to participate in the great battle near Tannenberg, the largest battle that this war had seen up to this time.

On that day we were located at the Sauerbaum Forest. At 10 a.m. we could see nothing, nor could we hear any shots, not even the distant thunder of guns that we had heard during the days before. All was calm, and none of us had any idea that we would be involved in heavy fighting in just a few hours. Suddenly the order came, "Mount!" And off we trotted. For a few kilometers through the forest we went at a steady pace among trees and shrubs. Paths didn't exist.

Suddenly, beside us the first shells hit. This was now my first time in combat. I cannot recall today what my thoughts were at that moment. The battery was in front; I was several hundred meters behind with our combat baggage and tried to reach the gun carriages. When I came to an opening in the forest we received fire and I had to have my men dismount and stay behind our baggage wagons.

No trace of our battery! That is why I sent our squadron farrier to inquire and report back to me about where the gun carriages had taken cover. He did not return. Later, I found that he had been badly wounded and transported to a hospital. Suddenly the call came for a stretcher-bearer. So my good, obedient, most of the time drunken stretcher-bearer Cibulski went on his way to answer the call. But just a few meters away from the baggage wagon he suffered the same fate. Four fellow soldiers carried the badly wounded man away. What a "nice" beginning that was for me. There was no one else, I was the only NCO and did not have the courage to send out another one of my men, because I had to assume that he too would be lost in all that heavy fire.

I did do the best one could in such a situation—I remained where I was and waited until the fire decreased somewhat. It was a terribly hot day. Our leaders, still very inexperienced, went through sandy soil on their route, and with each sandy cloud that appeared the Russians knew our location.

Because I was positioned at the only access road leading to the front line, I was in a really precarious situation. After enemy fire had ceased for half a minute, I had my men mount and in a full gallop we went to another spot that looked more suitable. To my great relief, there stood my cavalry sergeant with our gun carriages. If I had been slower in the approach to reach this spot there would have been very little left of us, as our enemy had noticed the dusty cloud that my baggage wagons raised and had aimed terrible fire at it.

I came to our gun limbers, where a first aid station for the infantry was being set up and for the first time saw the infinite cruelty of war. Over and over, newly wounded, friends and enemies, officers and rank and file were brought in. Horrible! The field dressing station only lasted at that spot for a quarter of an hour; it had to be broken up as our enemy was bombarding that particular area. None of us knew what really went on and whether the situation was good or bad for us.

Toward evening when it grew dark, the enemy fire decreased and when our gun limbers were pulled up to the guns, our captain, who had been to see the major, told us that it seemed we had won a major victory. Why and how this was done was very unclear to us.

I recall that we were told of a terrible situation that had happened that afternoon. One of our heavy artillery batteries was told to move forward to a position at a distance of about 5,000 meters [from the front line] and fire at our enemy. Meanwhile, our great infantry, storming ahead at a mighty pace, came in very close contact with the enemy. The net result was that the first salvo that our heavy artillery delivered fell on our own troops. What a depressing thought to be shot at from the front by the enemy and from the back by one's own troops. Thank God I have never been in that situation, but I can well imagine.

That evening we encamped. But first I transported a small group of slightly wounded to the dressing station because I was the one who knew its exact location. I saw hundreds and hundreds lying on the ground, men who had not yet been attended to because there were too few doctors at these first aid stations. Yet there couldn't possibly have been enough. Our battery had suffered no losses except for my two men. Only our Deputy Officer Glaubke had received a slight wound on one arm, which kept him away from the battery for a few days. He returned to us not yet completely healed.

Our troops who won the battle near Tannenberg were our elite, the advance storm troops who could not be held back for anything. We heard that the infantry officers, on the evening of the great storm that brought the decision near Sauerbaum, held back their men with revolvers in hand—they were going ahead with incredible rage. Would the same occur today, nine months later?

Very early the next morning we went on to pursue the enemy. We could see 200 captured [Russian] prisoners being led away, then 500, somewhere else 2,000, then again smaller groups, and at that point we came to realize that something big must have happened these last few days.

On the 27th we marched all day long; a few shots were fired here and there but we did not see the enemy. On the 28th we were close to Allenstein. In the morning an enemy plane on its way to inform Allenstein that our strong forces were approaching was shot down with our machine guns and so we were able to come within 2 kilometers of Allenstein without being noticed by the enemy.

A great many people [Russian soldiers], however, had already left Allenstein. As it turned out later, the Russians seemed to prefer dealing with our 17th Corps instead of us. Not many were left behind. A brigade was all that was still at Allenstein. We went into position on a hill right before Allenstein and our infantry fired at Allenstein. Our artillery was not allowed to shoot at all at that point.

It didn't take long before many Russians ran out along the two roads leading out of Allenstein. Now our artillery started. Both roads were taken under fire, causing a terrible massacre. Horses with smashed limbs, men with contorted expressions were lying along the roads, mortal fear in their faces. They had thrown themselves into the road ditches to avoid our bullets, yet they did not escape their fate. It was a terrible sight.

On 28 August, for the first time something gratifying happened to us. As we were moving into Allenstein, the population gave us a mighty cheer. People shared with us what they had: food, beer, wine, lemonade, handkerchiefs, anything we could use was handed to us to cheer us up and thank us. We needed that. After our exhausting efforts in those days of great heat it was like medicine for the soul to feel successful.

We left Allenstein that evening to encamp about 8 kilometers out of town. By 4 o'clock the next morning we started again, in forced marches, 2 hours of uninterrupted trot, then a 5-minute walk, followed by a 3/4-hour trot—and soon we knew that it wouldn't be long before we would be in combat. When we left the forest we saw men of a jäger battalion coming toward us in great alarm: they had used up all their ammunition and feared that they would have to withdraw and leave behind great quantities of Russian baggage that the Russians had been stubbornly defending. Ten minutes after we stepped in we won. Any of the Russians who could still do so fled in great haste and we were left with all their baggage: hundreds and hundreds of baggage wagons and 400,000 Rubles in cash.

Shortly before all this happened we had passed a farm near Wutrinen that had been raided by our infantry a day earlier. The Russians had been surprised while cooking, and we had seen men still sitting with their pots and pans in hand just as they were when the deadly bullets reached them. It was one of the most horrible battlefields that I have seen in the entire war. We camped that evening and were given the news that the enemy had been beaten soundly, and so we came to realize the full success of the battle of Tannenberg.

On the 31st we moved back to Allenstein and were left in peace until 2 September. On the 31st from Allenstein I wrote a letter to my wife, which I will put down here in its entirety except for personal comments:

> *Today we have a so-called day of rest, which means nobody is allowed to go anywhere, because there could, of course, be an alarm given at any time. I will use this time to write in more detail than I have done in my previous accounts.*
>
> *As you most likely read in the newspapers, we have done great deeds here, not only during the three days of heavy fighting but also during our 12- to 16-hour-long approaches, in which we*

fought smaller battles with the enemy's rear guards. And even afterward, when we thought that all was over, small enemy units appeared and shot at our men from farms and the forest, and so even then some good souls lost their lives.

Ours was a complete victory. Five enemy corps have been completely destroyed and were taken prisoner, and their baggage, which contained things of great value, was taken. Among other items there was a box that contained 400,000 Rubles in gold and silver. Hundreds and hundreds of caissons with important military items and plans came to us and were handed over. Other items were divvied up among officers and others. Out of principle, I never take anything and I follow my own rule strictly. Innumerable guns were found and other weapons, clothes, furnishings, even women's items—a sure sign that these dogs do not shy away from bringing along their women in such difficult times.

You might have gotten some idea on how I felt in battle when you read my postcard written on the 26th during combat. (I wrote then: "I am writing to you while the battle rages. We are in a terrible, bloody battle. The most awful thing I experienced, indescribable. We are achieving a great victory, but at what price, etc. ") It cannot be put into words how terrible all this is and how much one wishes that these gruesome battles would come to an end soon.

We are asked to do things that only the perfectly healthy can manage. Twenty-eight hours without food and 2 hours of sleep are the norm. If I wasn't so used to functioning with little sleep I would be more upset, but I am doing very well, thank God. The anxiety on approach into battle that many of us experience I do not feel. The first few minutes in battle though are unpleasant when shrapnel and shells hit nearby to our right and left while we get into firing position. But that soon ceases when combat is in full swing.

Afterward, however, when the wounded of the battle first appear, when screams and whimpers never stop, we realize what a huge, huge disaster we are in. I will give just a few details at this point; however, if I get to be an old man with gray hair I still will not forget any of this and 26 August, that determining day, will stay in my memory forever.

Good things happen too. On the evening of the battle just after we encamped (by the way, it has been 8 days today since I have taken my boots off), the news came that Allenstein had been occupied by the Russians on the 26th. They had proclaimed that the town and its surroundings were Russian and newspapers in German and Russian had been distributed by the Russians claiming that the German force was crushed. So we marched off at 2 a.m.

At one o'clock in the afternoon we were within 7 kilometers of Allenstein. News reached us at that point that many [Russians] had left Allenstein—all those who were then beaten decisively on the 29th—and that only one brigade remained in "Russian" Allenstein.

The [Russian] airplane that was to bring news to Allenstein of our approach had been shot down by our machine gunners that morning and therefore our artillery was able to take position within 2 kilometers of town while our infantry stormed it. When the deathly frightened Russians tried to flee out of town, we shot at them and caused a terrible massacre.

Afterward, our march into Allenstein was great! "We have our Prussians back," is what the crowd shouted and we were cheered on endlessly. Bread, beer, wine, lemonade, cigars, cigarettes etc.—the rich as well as the poor wanted to give us all they had. More than just a few little old ladies stood there with folded hands, praying. It was a rewarding, touching experience to look at these happy townsfolk.

And after that, on we went! A brief encampment, a straining march, and the massacre started over. In approaching [one of our jäger units] we could already see men, horses and caissons lying about in heaps and that our jäger unit was about to withdraw. But we arrived and within 5 minutes our "Jäger-Battery," a name we invented for the jägers in our 6th Battery, had beaten the enemy.

A badly wounded Russian was lying by the roadside and no one was taking care of him. He must have been there for quite a while before I got to him. I went to fetch a stretcher-bearer and a doctor. He was shot through the lung. On my request the doctor gave him morphine, he was bandaged and received something to drink. While blood ran from his lips he patted the stretcher-bearer, nodded toward me with a look that I will never forget, and at the same time gestured with his hands as if he wanted to say: "All this is not my fault."

I don't know whether he will make it, whether there is somebody at home who is worried about him. I only know of one who had big tears running down his cheeks while all around him there were shouts of joy—and while I am writing to you I have to swallow hard again to hold these tears. I am a soldier!—It is these minute details that embed in one's memory. However, instead, one is supposed to focus on the large goal only and ignore all else.

* * * *

Werner Riess had been involved in two of the largest battles in European history, Gumbinnen and Tannenberg, from 19 to 28 August. His recollections of those days reflected the confusion the soldiers experienced. Even seven months later, as he wrote this memoir from his field journal, he did not have a clear view of what had happened. It would be years before historians pieced together the German and Russian data to reconstruct these battles fully.

On 19 August the German Eighth Army had been entrenched near Gumbinnen to withstand a frontal assault by the Russian First Army. Riess's unit held the right or southern flank. The German generals expected the Russian attack in two or three days, so, as he noted at the end of chapter 1, Riess returned to the Angerburg supply depot to acquire more supplies for his battery. Then the whole situation changed rapidly.

Two days before that, German General François, holding the German left flank in the north, thought he saw an opportunity to destroy the Russian army piecemeal by attacking them first from the north before they reached the German defensive lines. Taking the initiative, he attacked and won the battle at Stallupönen, convincing General Prittwitz, the German Eighth Army commander, to order the German center and right flank forces to leave their defensive positions and attack the invading Russians. Thus Riess's battery quickly advanced into the Battle of Gumbinnen, leaving him behind near Angerburg.

The Battle of Gumbinnen of 200,000 invading Russians and 200,000 defending Germans was huge, confused, and bloody, flowing over many kilometers of land, but it ended as a stalemate.

The German troops on the right and left flanks did well, while the center had to retreat. Seeing this, and at the same time hearing that the Russian Second Army to the south was closing faster than expected, General Prittwitz at first ordered the entire German Eighth Army to retreat west, and to regroup near Nordenburg.

General Rennenkampf, commander of the Russian First Army, thought he had succeeded in his mission to hit the Germans hard and keep them occupied while the Russian Second Army advanced from the south. He therefore stopped his army where it was and readied his men to defend against another German attack.

Dennis Showalter, in *Tannenberg: Clash of Empires*, wrote that during the battle of Gumbinnen the rapidly firing 1st Reserve Corps' field artillery batteries almost ran out of ammunition. They had to change positions so many times to support the infantry's front lines that they lost contact with their ammunition supply platoons. That is what Riess recorded from his perspective. In command of 6th Battery's combat supply wagons, he never caught up with the guns during the confusion of their first combat and did not find them until the day after the battle, near Nordenburg. He must have been terribly frustrated at the time, but by April, when he wrote, he seemed to have accepted it as part of the confusion of war.

As the Germans withdrew farther west, 6th Battery was with the rear guard, repeatedly unlimbering (setting up their guns) to fire if needed, remounting and falling back, then unlimbering again. The German troops were hampered by tens of thousands of civilians clogging the roads. Families and whole villages filled their carts with belongings and children, drove their cattle alongside, and tried to escape the oncoming Russians. And yet 6th Battery traveled fifty to sixty kilometers each day.

The soldiers, and especially the horses from both countries, were suffering from days of forced marches and combat in that unusually hot and dry August. As the German forces retreated west, the field artillerymen were happy to come upon one of the army's stud farms in Gudwallen, where for some reason the staff had not evacuated hundreds of horses. The farm had been hit by Russian long-range heavy artillery, scattering most of the horses. The artillerymen euthanized the wounded horses, took the best they could capture for their needs, and gave the rest to the fleeing civilians in order to deny the Russians fresh horses.

The German High Command at headquarters in central Germany, also confused about what was happening on the Eastern Front, thought General Prittwitz had lost his nerve and therefore replaced him with retired General Hindenburg. The original plan had been to destroy the Russian First Army before attacking their Second Army, but there was no time to destroy the Russians at Gumbinnen. Instead, Hindenburg left a small force facing the Russian First Army to impersonate the entire German Eighth Army, then shifted most of the Eighth Army south to face the advancing Russian Second Army.

Germany's 1st Reserve Corps was to remain in place near the Masurian Lakes to fight against the Russian First Army's left flank or the Russian Second Army's right flank, depending on how the battles proceeded. Thus Riess recorded that on 23 August his battery was told to stay where they were for a few days. He did not know why. To his west, unknown to him and his unit, there transpired a masterpiece of military logistics and maneuvering. The Germans used railroads and forced marches to swing some 100,000 men quickly through East Prussia to face southeast, with all their artillery, horses, wagons, and supplies.. And so, near the town of Tannenberg, the approximately 170,000 Germans who were not casualties in the Battle of Gumbinnen faced the 200,000 men of the newly arriving Russian Second Army.

Led by General Samsonov, the Russian Second Army had been on a forced march through sand for several days in hot, dry weather. They were hungry and exhausted but on schedule. Samsonov had a wide reputation as an able and brave officer, with a sterling career behind him as a Russian cavalry officer in many battles. Unfortunately for the Russians, they did not have a telephone line between their First and Second Armies, meaning that they had to communicate by radio, and the Germans were listening. The Germans knew where the Russian armies were, while Samsonov did not know that most of the German Eighth Army was maneuvering into formation ahead of him.

When the Battle of Tannenberg began, the German middle (20th Corps) fell back when the Russians attacked, luring General Samsonov into a trap. The German left and right flanks planned to encircle Samsonov's army. Riess's battery was to be part of the northern trap door to help confine and destroy the Russians.

As 6th Battery moved southeast, Riess had his first experience of being a target of the Russian artillery. He related to his wife later that the first shells exploded close to him, startling his horse Intarsia. She took the bit in her teeth before he reacted, and bolted out of the forest, straight for the Russian lines. It took him a few seconds to turn her head and steer her back to his men.

North of the main battle, but south of Riess's position, a Russian battalion had taken and held Allenstein, threatening the Germans' left flank. The 36th Reserve Division, including Riess, was sent to retake Allenstein as quickly as possible. On the way, soldiers in his regiment shot down a Russian scout plane that happened to be carrying a message to Allenstein's commander, warning of the coming Germans and ordering him to put the torch to the city and march to the aid of another Russian unit. Luckily for the German soldiers and the city's inhabitants, the message didn't arrive.

It is interesting to note that as the German 36th Reserve Division approached, the commander, so as not to destroy the town, had his infantry attack Allenstein without artillery support. He let the artillery fire on the Russians only once they were retreating, clear of the city. Meanwhile, the division almost had the city surrounded, with 6th Battery covering the east road. As the Russians poured out of Allenstein in retreat, the battery's four guns opened up, killing and wounding many of them along the east road.

The battle for Allenstein was numerically a small part of the Battle of Tannenberg. However, it was important enough to catch the attention of most historians writing about the Eastern Front and at least two notable novelists—Alexander Solzhenitsyn in *August 1914* and Ken Follett in *Fall of Giants.*

Immediately after taking the city, 6th Battery was ordered south again to help close the encirclement of the Russian Second Army. As part of that, they were ordered to support two jäger battalions that were pinned down by Russian machine guns while trying to take a small town. There Captain Krüger, commander of 6th Battery, won accolades for his "brilliant art in estimating distances," so that the first three salvos landed precisely in the Russians' infantry forward line.[3] As the Russians began to retreat, their machine gunners prevented the German jägers from advancing. Captain Krüger then moved the battery closer and put the machine gunners to rout, allowing the jägers to capture the town, including the Russian Army's entire baggage train of more than 500 wagons, with its huge cache of money in gold and silver coins.

After securing the war booty and burying their dead, the field artillery men and jägers bivouacked (camped without tents) together that night, celebrating not only their immediate victory but that of their whole army at Tannenberg. The battery returned to Allenstein the next morning for a few days of regrouping and maintenance, as the rest of the German Eighth Army marched up from the south.

[3] *Das Reserve-Feldartillerie*, p. 15.

While they and the infantry held Allenstein, a train brought them their first mail and packages from home. The field artillery men also celebrated the news that elsewhere, the rest of 2nd Battalion had captured some Russian kitchen wagons, so that finally, after three weeks, each battery would be able to have hot meals in the field.

We can see that several months later, when he wrote this memoir, Riess still did not know that he was part of the northern trap door that helped destroy the Russian Second Army. After three days of fighting around Tannenberg, only approximately 10,000 of the 200,000 Russians in the Second Army had been able to escape. Almost all had been killed or captured, along with the capture of most of their weapons, ammunition, ambulances, kitchen wagons, supplies, and horses. General Samsonov, the Russian commander, committed suicide.

Hearing the news, General Rennenkampf, who had planned to attack west with his Russian First Army to capture Königsberg, withdrew east. He assumed the Germans would take a few days to reorganize after the Battle of Tannenberg. However, Hindenburg and his officers were indoctrinated with the advantages of continuous aggressive action. Without pausing, the German Eighth Army marched northeast to destroy the Russian First Army, or at least push it out of German territory.

Map 3. 2–26 September 1914. Allenstein to Gielgudiczky & Eydtkuhnen. Battle of the Masurian Lakes.

CHAPTER 3

Battle of the Masurian Lakes and into Russia

2-26 September 1914

One month of war has gone by and after initial failures we have now done extremely well. And on we go without stopping, always forward, forward, forward. From Allenstein to Wartenburg, where we had a day of rest, decent quarters and a bed, a real honest to God bed! From Wartenburg we went to Lisettenhof, where I became aware again that I have gallstones.

From Lisettenhof we went to Mathildenhof, then to Rotkerpen, from there to Dietrichsdorf and then to Arnsdorf. Beyond Arnsdorf near Gerdauen the Russians had moved into very strong positions. No need to say much about our rate of marching; one look at a map reveals what tremendous distances we covered.

On 2 September we had left Allenstein and we arrived at Gerdauen on the 8th. Gerdauen, to which most of the defeated Russians had withdrawn, is situated on a hill. The enemy had left behind smaller units to prevent us from making too much headway.

The movement of the main troop was easily discernable. When they first reached a village they would immediately set fire to it, so that the rear guard troops would know where the Russians were located and where they had to follow.

By dusk we were close to Gerdauen and settled into very secure positions as there certainly was going to be heavy action. We came under incredibly heavy artillery fire. On the morning of the 10th we were to start storming Gerdauen, but there wasn't a trace left of the Russians. They had left Gerdauen in the middle of the night and we had to follow them.

What, however, had become of Gerdauen?! We had marched through this prospering town before and now we saw a miserable pile of rubble. Nothing was left standing. A house here and there, otherwise just walls, and the Russians fortified behind these walls. The fortifications that they had built in town as well as along the roads were exemplary. And the storming of Gerdauen would probably have been a terrible massacre as it was so greatly fortified.

Our unsurpassable Hindenburg had made the plan though, that our 17th Corps was to attack Gerdauen from the opposite side and the Russians had learned that lesson at the battle of Tannenberg. Better to run than to end up in the stew!

Well, as I said, they were gone. So on our way we went. I mentioned the towns through which we marched when I described our way from Allenstein to Gerdauen and I will do the same now. From Neuhof we went to Friedrichswalde, then to Melchersdorf, where we took quarters for a mere two hours, and on we went to encamp in the evening. In Melchersdorf our captain received the Iron Cross. We were all incredibly proud of him. The next day via Bokellen (where the railroad bridge had been blown up) we went to Harpenthal, a small village close to Insterburg. There we got into some heavier action.

I want to mention that during all these days there had of course always been smaller fights. Near Harpenthal my battery had been assigned to a cavalry division, which meant that we marched at the front and had to go forward during an attack that was planned along a road near Insterburg. Beyond the road heavy enemy artillery had taken position and was able to observe us extremely well. It is a miracle that we didn't suffer losses. My dear old horse Intarsia, on which I was sitting, was shot badly. Things could have gone differently for me! It's one of the memorable days of the war for me, that 11th of September. I rode that horse for another 8 days, despite its great sufferings, then it was transported into a better world. Honor to its memory!

Once in a while among all that misery there were happy hours as well. The happiest of all was our march into Insterburg. No, not the happiest of all–that was my re-entry into Berlin. In my memory this march into Insterburg was just great, something I will surely never forget. Insterburg had been occupied by the Russians for 3 weeks and had been headquarters for General Rennenkampf and Grand Duke Nikolnjewitsch. Half an hour before we entered Insterburg those two dirty dogs ran off. Too bad! They would have made a worthy prize!

It is hard to imagine the enthusiasm and joy with which we were welcomed at Insterburg. The formerly mentioned great welcome we received at Allenstein pales in comparison. People were overjoyed. They threw so many flowers at us that we had a hard time passing with our horses and weapons. The shouting and enthusiasm of the crowd was indescribably exciting to us, and when the church bells started ringing around 8 p.m., a signal that Insterburg was German again, even the toughest among us had tears in their eyes. The city was illuminated that evening and people were so overjoyed, they didn't know how else to please us.

We had hoped to get a bit of rest, but when I reflect on this today, I realize that that was impossible. There were many fronts we had to fight against. Even the smallest unit was needed. Even if we cursed once in a while and thought this could not go on any longer, it did go on because it had to—it was impossible to let the troops rest. It is interesting to trace our steps on a map, the headway we made from 2 to 11 September, taking into consideration that for 2 of those days we were in the battle near Gerdauen.

The next morning we continued to pursue the enemy. We bivouacked near Malwischken. The nice weather had gone by and it poured terribly. In the last few days we took 30,000 prisoners and took away a great deal of war materiel.

It is easy to march during nice weather; marching in the rain is terrible, much more terrible than marching in snow, something we were to experience later. In addition, there was this unbelievable tiredness. I had slept no more than 5 hours in the last 72 hours. In

Inglauden, which we reached on the 13th, we had our first thorough sleep since Allenstein, because we rested until 2 p.m. on the 14th.

On the 14th at 5:07 in the afternoon we crossed the Russian border near Slowiki. Only now did we start appreciating the good German roads we had left behind. We got into unbelievably terrible mud. Up to their knees our horses went in, yet we had to move on; the rumbling of guns we heard told us that we were close to the enemy. At 7 p.m. we were in heavy combat again and at 10 o'clock we encamped in bitter cold weather and rain.

Earlier on that evening my first large supply wagon had fallen into the road ditch, if you can call it a road at all, and with the aid of tree trunks, which had to be felled first, with enormous effort we pulled it out again. Because of this, I got to the battery much later and received a great welcome, as I had generally been assumed dead.

Not a great night, it was our first cold night. In my diary I wrote about a freezing cold night. What real freezing cold conditions were like, I learned only later in those beautiful nights in November and December when we were encamped. Yet I don't think we really have any idea what hardship our troops suffered in February in the Carpathian Mountains in terms of cold temperatures.

The next morning the always bravely withdrawing Russians had disappeared and, most offensive to us, without ammunition had taken [their] 50 guns across the river Memel. I think those 50 guns would have suited us well, but that was all done with. Farther along those terrible roads we went to Gielgudiczky. On the 16th and 17th we rested. We slept in the barn of a farm property that was located directly along the Memel and offered a great view of the river valley (I had chosen the barn over the house and that was a good choice; those few who had chosen the house received unpleasant visitors [lice]).

On the 18th we were going back to Germany. On we went along terrible roads during a bad storm for 13 hours without stopping once. It was one of the worst days we had endured until then. That evening at least we stayed in a warm apartment with Germans who took us in kindly. We didn't take off our boots, though. It would have been impossible for us to put them back on the next morning as they were soaked with water. Off we went the next morning via the almost completely destroyed Eydtkuhnen to Wirballen. It was still raining hard, but at least we went along German roads until we reached Eydtkuhnen. In Wirballen we led a very pleasant life. We cooked and fried up a storm to recover from all that we had suffered and endured recently.

Sadly, my old Cavalry Sergeant Martin, whom I had befriended, we lost due to an illness and he did not return to us later. My new sergeant was Sergeant Schlieter, an honorable replacement in every aspect.

I will now add part of a letter I wrote home from Wirballen on 21 September:

> *What we have been going through here is hard to put into words. Our commander-in-chief has released the following army field order (as preface to what I want to describe to you later).*
>
> "Army field order of 15 September 1914: Soldiers of the 8th Army! You have wound new laurels around your flags. In a two-day battle and several days of ruthless pursuit through Lithuania and far

beyond the Russian border you have not only beaten the new enemy army consisting of the [Russian] 2nd, 3rd, 4th, 20th, 22nd Army Corps, the 7th Army Corps, the 1st, 5th Rifle Brigade, the 53rd, 54th, 56th, 57th, 72nd, 76th Reserve Divisions, the 1st, 2nd Cavalry Guards Division and the 1st, 2nd, 3rd Cavalry Divisions of the Wilna Army, but you have crushed it. Procured from these large battle grounds have been several flags, 30,000 uninjured prisoners, at least 150 guns, many machine guns and ammunition trains as well as countless war wagons. More war booty is still being procured. Thanks to your eagerness to fight, to your admirable marching performance and your shining courage. Honor God, he will continue to be with us. Hail to His Majesty the Emperor and King.

Commander-in-Chief
Von Hindenburg,
Major-General"

I underlined the "admirable marching performance" as this was one of the most brutal series of marches accomplished in any war. (Shortly past Wirballen it was to get much worse . . .) Two days in Russia; roads with giant, meter-deep holes, 18 hours in the saddle without once dismounting. Add to this rain and storms of such strength that you thought you and your horse would be thrown over. We were so used to hardship, yet at times we shed tears of fatigue, especially since we could see that so many men and horses were half dead from over-exertion.

At times I thought that I couldn't go on. Water ran down my back. Despite a coat that was covering my boots, there was water accumulating in these boots, and rain, hail and gales lashed into my face. Yet a man can endure more than he thinks possible. And endure I did. Half the battery had a bad cold; wonderfully, I did not, in spite of the wet boots that I could not slip off my feet.

Now we are at Wirballen, a genuine Russian town, in genuine Russian filth. But we are at rest, a well-earned rest. Sometimes I asked myself what awful sins I must have committed that I needed to pay so terribly. And pay I did. In those days one got to know God, who must feel terribly kind, very, very kind to let us survive this.

At least I always got to eat bread and butter with cheese, which my excellent Sell [Riess's orderly] handed to me on my horse. "Corporal, Sir, you need to eat!" When we came into quarters, he made tea for me and if possible, he prepared eggs, and he serves me well in every way possible.

The whole division envies me Private Sell and all the mail I receive. Right now, he is busy in the kitchen, preparing rump steak and potatoes to which my cavalry sergeant (who is very nice and whose good friend I am) and some others have invited themselves. He spruced up the house I presently occupy, which had been left by its owners and plundered by Russians and Germans.

Now I can lie on a couch, a hard and bent one, mind you, and one so filthy that we have covered it with a horse blanket and use a second one to cover me. Then a little fire gets lighted, a cup of tea brewed, and all the misery that has happened is forgotten—one lives like a gentleman.

Close to Wirballen we worked on entrenchment, which, as we would find out later, served us well. On the 22nd I received news of my advancement to cavalry sergeant. I am the first one in my division to be promoted.

****

Sergeant Riess, and everyone else in the German Eighth Army, had a right to be proud of their success at pushing the Russians out of East Prussia. During the Battle of Tannenberg, Russian General Rennenkampf began to shift his Russian First Army south to help the Russian Second Army, but then discovered that it was too late to take part in the fighting. Upon hearing that the Germans had already defeated the Second Army, General Rennenkampf stretched his army to form a long defensive line from the Königsberg area, then southeast past the northern tip of the Masurian Lakes to the German-Russian border.

The Russians were dug in, reinforced with several divisions, and resupplied, and they waited for the Germans to attack from the southwest. When they heard that the Germans had actually destroyed almost all of the Second Army near Tannenberg, the Russians knew that they were about to be attacked by everything the Germans had on the Eastern Front, but the Russian Tenth Army was coming to help from the southeast. If General Rennenkampf's First Army could hold its defensive line until their Tenth Army arrived, the Russians would have the Germans outnumbered two to one and surrounded.

At the Battle of Tannenberg the Germans had captured, wounded, or killed more than 90 percent of the 200,000 men in the Russian Second Army and captured most of their artillery, machine guns, and supply wagons. Only 10,000–15,000 Russians escaped. General Hindenburg then quickly swung the German Eighth Army northeast to destroy the Russian First Army. At the same time, he ordered newly arriving German divisions to shift farther north to attack the enemy's right flank, and he sent some of his best troops between the Masurian Lakes and south around them to try to encircle the Russians.

But General Rennenkampf was not about to be caught in the obvious trap. He had the Russian frontline veteran troops construct formidable defenses and wait for the Germans on a line from Gerdauen to Angerburg, with several newly arrived divisions positioned to cover their right and left flanks. Corporal Riess was at Gerdauen, west of the center of Hindenburg's Army in this third intensive engagement of approximately half a million soldiers, now called "The First Battle of the Masurian Lakes."

6th Battery faced the Russian lines at Gerdauen and immediately they began a field artillery duel while the infantry on both sides dug in and kept each other at bay. Both sides' field artillery units were rapidly mobile so that they could fire a few rounds and quickly move to new positions. The Russians had the advantages of more time to construct defenses and occupying a town that sat on a hill. The latter gave them better range for their guns, a better effect for their shells upon the Germans below, and a better position for observing the Germans.

For two days a low-lying fog kept the Germans from targeting the Russian positions accurately. Meanwhile, a tall bell tower rising above the fog in the middle of Gerdauen, beyond the maximum range of the German guns, allowed Russian observers in the tower to spot the Germans as they shifted their guns. For forty hours Riess was kept busy moving 6th Battery's field supply caissons and wagons wherever the unit's four guns would shift, supplying the gun crews with ammunition, food, and maintenance supplies, while sending teams back to the regiment's supply column to replenish his wagons. His experience as a department store manager may have served him well.

The German field artillery, augmented by a unit of heavy howitzers, tried many times to hit the bell tower in order to deny it to the Russian artillery observers, but the Russian heavy artillery kept them just out of range of that tower. On the second day of fighting at Gerdauen, 6th Battery was situated just north of the town when the fog lifted enough for them to close in for two shots. The battery dashed in at a gallop, unlimbered close to Gerdauen, and Captain Krüger used their precision targeting optics to aim the four guns quickly. With two shots each, they punched several holes in the bell tower and galloped back out of range. Evidently the Russian artillery observers abandoned the tower, as their artillery fire after that was much less accurate. Krüger would be awarded the Iron Cross for this dangerous action.

Even without the Russian observers using the tower, the Germans would have suffered heavy casualties in Hindenburg's plan of attack, as they were to storm the Russians' earthworks and barricades in and near Gerdauen while other German divisions would sweep around the Russian flanks. However, when the Germans attacked and soundly defeated the Russians' flanks in the south near the Masurian Lakes, the Russians abandoned their defensive position and withdrew to a new defensive line closer to the Russian border. Thus Riess's unit was spared what probably would have been serious carnage during the assault on the Russians at Gerdauen.

The Germans entered the town on 10 September to find it consisting of rubble, from both the German artillery fire and the Russians' use of material from the buildings to construct extensive and strong defensive positions. Neither Riess nor the official regimental history mentions what happened to the inhabitants of Gerdauen and most other towns and villages that were occupied and destroyed.

The German soldiers were concerned by the expert fortifications the Russians had built in just a few days; the Russians evidently had learned much about constructing defenses against modern weaponry during the Russo-Japanese War in the Far East a decade earlier.

General Hindenburg's men, not resting half a day, continued a fighting advance against the Russian rear guard, who worked hard to slow down the Germans as the rest of the Russian First Army created strong defensive positions in a new line centered on the city of Insterburg. As they withdrew, the Russian rearguard repeatedly unlimbered iletheir field artillery and their protecting infantry deployed. They fired a few artillery, machine gun, and sniper rounds toward the pursuing Germans. This forced the Germans to stop and deploy because they could not see whether they were finally up against the Russians' major defensive line or just the rear guard.

The Russians then would limber their field guns and march north again a few kilometers, repeating the routine whenever they found a good position. With this tactic they slowed the pursuing Germans all along the way, so that most of their army could reach the new major defensive positions in safety and dig new emplacements.

It was during one of these many delaying battles, on 11 September, that 6th Battery was assigned to the division's hussars. German hussars were light cavalry, typically used as troop-strength scouts or for attacking lightly defended targets with 3-meter steel lances. Even in the modern warfare of the First World War, hussars were particularly effective against artillery as they could sprint across a battlefield if no major tangles of barbed wire had been deployed. They galloped abreast with their lances aimed at the artillerymen, who might have time for a few quick shots unless they were so terrified that they froze in place or fled.

At 11:45 a.m. 400 hussars, with 6th Battery just behind them, swept around the Russian infantry's right flank at a full gallop and charged an artillery position to the west of Insterburg. The German field artillerymen hoped to find a raised position where they would quickly unlimber to fire one or two rounds into the defenders, possibly to destroy machine gun crews or other high-value defensive targets. Riess and his supply team would have been galloping just meters behind the battery's four guns. During the battle the Russians did not retreat, but fought. A Russian bullet, possibly meant for Riess, hit his horse Intarsia and began her slow death. Riding her for the next eight days, "despite her great suffering," must have been traumatic for him.

The hussars successfully destroyed the enemy artillery position and everyone who survived galloped back. That evening the regiment marched into Insterburg, to be lauded by its Prussian population, who had been the involuntary hosts of General Rennenkampf and his staff since the beginning of the war.

In August the Russians had quickly taken Insterburg and established their First Army headquarters in Dessau Courthouse. Early in September Grand Duke Nikolai Nikolaevich (Nikolnjewitsch in Riess's memoir), commander of all the Russian armies on the front, was visiting General Rennenkampf in Insterburg for a few days before the Germans reached the city. The two generals escaped entrapment by just a few hours, Rennenkampf leaving at the last moment.

Unknown to the Germans, during Nikolaevich's short visit, he and General Rennenkampf had developed a new plan to match their circumstances. The Germans' Eighth Army had completely destroyed the Russian Second Army and was pursuing their First Army. Now, the Russian Tenth Army, made up of some of Russia's best troops, was on its way to the front but hadn't yet arrived. The Russian leaders therefore decided it was best to put off any more major battles with the Germans for a week or two. Rennenkampf ordered most of the Russian First Army's slower moving units, such as heavy artillery, hospitals, and supply convoys, to march northeast. He ordered the infantry to hold the front and make small attacks to cover the slower units' retreat—he planned for a well-organized fighting withdrawal east, into Russian territory.

Meanwhile, the German Eighth Army, reinforced to its original strength of 200,000 men, remained an orderly fighting unit spread along an extensive front. General Hindenburg kept his men methodically advancing on the Russians, in order to push them out of German territory. In and around Insterburg, Riess's battery slept for a few hours, the first in days, but were already up at 3 a.m. By 4:30 a.m. they were marching quickly east in pursuit of the Russians, trying to capture some of the rear guard. Toward the end of the day they finally ran up against the strong new Russian defense line at the village of Malwischken.

In hot pursuit, the 36th Reserve Division commander had found it best to attack the retreating enemy quickly when they stopped. He employed the infantry, hussars, and field guns simultaneously to try to overrun the Russians before they could construct more than rudimentary defenses. However, here the Russians were already prepared.

At Malwischken, 6th Battery was to the northwest of the town, where they advanced on the forest but found no enemy. However, part of the battalion, after fierce fighting, captured the town, many prisoners, some supply wagons, and a Russian field kitchen with warm food still in it. That night the regiment bivouacked in pouring rain but at least with a hot meal. In the early morning they again pursued Rennenkampf's army, fighting and capturing a rear guard unit at Pillkallen.

These five days of intense fighting (7–13 September) are now known as the First Battle of the Masurian Lakes, though the battle only started there and moved far to the northeast. Sources differ, but the battle costs were high. The German Eighth Army lost 40,000 to 70,000 men. The Russians suffered the loss of 100,000 to 145,000 men.

6th Battery rested most of the next day, 14 September, while the division began to cross the Szeszupa River into Russian territory near the town of Slowiki. Late in the afternoon 6th Battery crossed the river while a military band played music to keep their morale high, then they fought their way north.

In addition to confronting the Russian defensive lines and ambushes along the way, here and elsewhere the men and horses of 6th Battery had to contend with Russian roads turned into quagmires by the recent heavy rains. The men often dismounted to walk beside their guns, caissons, and wagons and help extricate them from the mud. While the German roads were mostly dirt covering a solid base of rock or gravel, the Russians had purposely left the roads on their side of the border without a solid base. When it rained and in the spring the Russian roads were all but impossible to use. Their condition was part of the Russians' long-term strategy to slow down any invading European army. It proved effective during and after the severe rainstorm of 14 September that Riess described.

As the Germans passed through the Russian towns east of the Szeszupa River, Russian Jews often greeted them with gifts and invited the officers into their homes for tea. Many Russian Jews saw the Germans soldiers as liberators from the czar. To help consolidate their extensive empire of various cultures, the Russian czars, encouraged by their religious leaders, severely mistreated anyone who did not convert to the Russian Orthodox religion or leave the empire. The severity of abuses had increased during recent decades, especially involving Jews. In comparison, in 1914 the Germans officially restricted rather than overtly oppressed non-Christians.

The next morning, Riess was on the south bank of the Memel (also called the Niemen or Neman) River, over which the last Russians had escaped with their artillery. Riess marched with his battery east along the river to Gielgudiczky. For two days 6th Battery remained on the south bank of the river, facing the Russian First Army. However, Hindenburg received word that the Russian Tenth Army, new to the campaign, was marching to attack his army from the rear and right flank; hence he ordered a withdrawal, back into Germany.

After returning to German territory, the Eighth Army deployed along the border, facing east. 6th Battery marched south to Eydtkuhnen, an important railroad town near the border. Then they marched east, again across the Russian border, to the nearby Russian town of Wirballen, where they deployed, entrenched, and were able to rest and resupply for a week. In Wirballen someone, possibly Riess, had to put down his horse Intarsia, because of her suffering from the bullet she had taken eight days before. It must have been hard on him. He then rode his other horse, Fuchsrot. It was during this break in the fighting that he received his promotion to cavalry sergeant of the Prussian Army, a coveted rank for a German soldier.

While Riess and the other men of 6th Battery wondered why they had given up the chase and withdrawn from the Russian territory they had captured with their toil and blood, Hindenburg and his staff knew that the German Eighth Army had been in a precarious position that easily could have become a disaster. They could not expect much reinforcement from the west, as the western German armies were locked in battle with France and her allies.

With the arrival of their Tenth Army, the Russians would again outnumber the Germans two to one on the Eastern Front. The Russian Tenth Army approaching the German border consisted of approximately 200,000 fresh, well-trained and well-equipped regular troops, including many divisions of the highly respected mounted Cossacks and equally respected Siberian infantry.

General Hindenburg's imperative was to ensure that everyone and everything in the German Eighth Army was in order and prepared to fight the next onslaught he expected from the east.

Map 4. 27 September to 10 October, 1914. Eydtkuhnen to Augustow to Eydtkuhnen.

CHAPTER 4

Feigning a Solid Front

27 September–10 October 1914

The weather had been good, but from our past experience, we could well imagine that it would change the day we had to march off. And when the signal was given on 27 [September], the beautifully blue sky we had enjoyed until then changed into thick, heavy rain clouds. It didn't take long until the rain started, and when we arrived at Mariampol from Wirballen in the evening we were already soaked to the skin. At least we had made rapid progress because the road, even for our needs, was quite passable.

Rumor had it that enemy troops had accumulated around the Augustow area for the purpose of invading East Prussia, and we were to prevent this. We marched accordingly. On the 28th we were supposed to get to Suwalki from Mariampol. It was simply impossible. Rain and gales raged as never before in this campaign. Therefore, around 6 p.m. we had to stop in Andrejewo, a small village, if you can even call it that—it was really tiny. 40 of us men stayed in a room that belonged to a family with a large number of offspring. "Soaked through" is a mild way to describe our condition at that point. Our poor horses and some of our men had to encamp in this freezing cold rain and a few, who later had to go on sick leave because of rheumatism and related problems, probably were afflicted with these ailments that night.

The next morning we went from Suwalki to Raczki. I wrote in my diary: "This is even more exhausting! Pounding rains and gales as never experienced." It was unbelievably brutal. That day the rain didn't stop for one second; it always hit us from the front, and we were fighting against the gales. At least we had nice quarters at Raczki. I was able to lie down on a couch, take off my clothes, had some warm tea, and frankly was almost able to forget all the exertions of the preceding days. What people can endure when they have to!

We weren't able to enjoy our pleasant quarters very long, because on the 30th we went from Raczki to Janowka early in the morning, and there we went into field emplacement. We waited for the enemy, who did not show up, and by midday we changed positions. We found a new one very close to Wielgi-Pruski [Pruska-Wielka]. The night remained peaceful; we slept in a hut together with its sick inhabitants, who were terribly afraid of us, the barbarians, who had given them cash and something to eat. Only when they realized that we weren't as bad as the Russians had made us out to be did they become friendlier. Next morning the shooting started. By midmorning we already had two injured men. The weather stayed the way it had been in the days before, and our success was as poor as the weather.

On 2 October we were in the process of getting into a new position but the Russians had taken notice of our re-positioning and taken us under heavy fire; we had to withdraw rapidly and got back to our old position unscathed.

That evening we returned to our former quarters near Wielgi-Pruski, where a medical doctor who had moved in was very displeased that I wanted to lay my weary head to rest at the place where he and his patients were staying. I tried to persuade him and after a while he gave me permission. We became great friends. As it turned out, he was a friend of Dr. Domnauer and Dr. Meyersohn and we reminisced about them. Unfortunately, he is in a Russian prison camp today, as he was captured during the heavy battles near Prassnitz in January; the same battles that were very costly to so many brave men in my battery. But I myself was no longer with them at that time. We ate and drank together all we had and slept among the wounded. When I woke up the next morning, the man lying next to me was dead. What strange sleeping partners one experiences at times like this!

How the battle was going none of us knew. Once it was said to be going well, then not so well. On 3 October we knew for sure—we were to regroup. We all know what that really means: immediate withdrawal! Our left wing had been victorious, but the right wing, the one where we stood, did very badly. The leader of our corps, Artillery General von Schubert, was reassigned from our corps at that time. Well, we all had to be "re-grouped."

I would like to take this opportunity to point out that our cavalry did a downright infamous job in all these battles from the very beginning. To those of you who read these notes and disagree, I want to make one concession—I claim that at least the cavalry allotted to our 1st Reserve Corps [the 1st Cavalry Division] did an infamous job. Nowhere was this more apparent than during our withdrawal, the withdrawal that was about to happen.

The weather turned nice again, a sure sign that everything would be peaceful for a while, and on 3 October in the afternoon we arrived in Niedzwietzken. Since the Russians had not even dreamed of pursuing us (as, according to our division headquarters, they themselves had suffered great losses), we were able to rest. On the 4th in the morning I had planned to go to Marggrabowa, located about 7 kilometers away, when suddenly shells landed in our ordnance park [where the guns were emplaced]. Our cavalry had scouted so brilliantly that the quarters where our division was located had Cossacks [fierce Russian heavy cavalry] in them.

At that point I saw for the first time how Prussians can withdraw in utmost haste. Our train [convoy] of supplies that had been approaching escaped down the road in a wild chase, and if the Russians had proceeded more cleverly and had waited just half an hour before opening fire—who knows what would have happened to us?! But as it was, we went into firing position and after 1½ hours had beaten back the Russians. In my diary I wrote: "Most dangerous situation in this war until now." Good thing I only wrote "until now," because later a lot more would happen. At this point I was made leader of my squadron.

That evening we left Niedzwietzken during a lovely warm fall night, and we moved via Marggrabowa (which was completely intact at that time) to Schareyken, which we reached by midnight. We were given quarters at some nice people's home (later on everything was to be taken away from them). I asked the daughter, who was about to travel to Berlin, if she could deliver greetings to my wife, which she faithfully did. I don't think that she regretted it.

The next morning at 4:30 a.m. we went on to Goldap, where we boarded a train. Very appropriately, we were transported in cattle cars; we didn't feel much better than that after all we had endured and been through. We went to Eydtkuhnen. Our quarters that night were located in Absteinen. The sky was crimson as a huge battle was in progress. All night long there was heavy firing. We had been assigned to be a reserve unit.

I mentioned earlier that we had dug entrenchments near Wirballen. Now the corps that was engaged in heavy fighting took over these secure positions, and they were almost invincible. At 5 the next morning, after passing Stallupönen we had to move to the right side of our position, because the enemy was trying to get around us. It seems that the enemy tired of trying to do so, and therefore that evening we were able to go back to Eydtkuhnen to our quarters, where we stayed until the 7th, constantly on alert. At least a wagon that had arrived filled with wine and other pleasant surprises got us into a better mood for one day. I wrote in my diary, "General drunkenness!"

Alarm on the 8th! From Bilderweilschen we went to Barzkehmen then to Kosakweilschen, where we encamped. The next day, 9 October, we saw combat near Schirwindt. Victory along the entire line! At one point the enemy had shot at our limbers to such an extent that we lost several horses and one NCO was injured. At night we took quarters in some very drafty barns.

An eerie sight was that of the barns that were burning in the town of Neustadt, through which our infantry was storming. Our battery was positioned on a hill and supported the attack, and it was horrid as well as uplifting to see how our brave men went ahead.

On 10 October we went back to Eydtkuhnen. I will add here the contents of the letter [to Gertrud] I wrote on 10 October:

> *Where to begin and how to write everything that has happened to me the last 14 days since my last detailed letter, I do not know, because it would take day and night. And I know, that's not what you want me to do, as I am extremely tired right now.*
>
> *Since the alarm that was given on the 8th and of which I wrote to you, we have been in fierce battles again. The enemy facing us has been thoroughly defeated near Schirwindt, yet that is only one of the battle lines of the whole battle near Wirballen.*
>
> *As I already told you, it will most likely be one of the biggest battles of this war (not in the least true, as it turned out). At the enemy's left flank and especially in the center, our brave troops have been battling for 7 days and if things don't go well—let's hope this won't happen—we'll soon get another alarm and off we will go! That's why we can't really call this rest; we are constantly sitting on top of a volcano.*
>
> *Let me tell you briefly what has been going on in these eventful last 14 days and what has stayed in my memory most clearly. One of the most terrible of*

terrible things in my mind was the weather that we had to endure between 27 September and 3 October. I don't know how to describe it. Rain and gales were in full force. Down my coat the water ran into the leg of my boots to collect at the bottom. We couldn't stay on our horses long as the gales almost knocked us over. All day long not a chance to empty the boots.

On and on we went for 16 to 17 hours, only interrupted by combat, and at night in a small room together with 40 men. If that wasn't to our liking we could go out and bivouac in order to live life to the fullest. I did not have that kind of pride and we preferred to move in together, 40 men in dripping clothes, almost frozen to death to get up at dawn the next morning to mount the horses and possibly outdo the exhaustion of the day before. And we managed to do just that! The worst day was 29 September, when we moved from Andrejewo via Suwalki to Raczki. It was the most exhausting of all.

That night was great, though. I was able to celebrate an anniversary that my dear Sell had figured out for me. Exactly four weeks ago in Wartenburg was the last time I had had a bed to sleep in and now I was to enjoy one again. I was in it from 9 p.m. to 4 a.m. At first I didn't know what to do with the down cover but to this day I relish the thought of a real bed. In those days, it was the only time I took off some of my clothes.

The next morning we went into battle. There was heavy fire and two poor fellows in our battery were wounded. The following day too, there was heavy fire. When we tried to change our position that evening we came under extremely heavy fire and had to stay where we were.

In my quarters, if you can call filthy Russian huts that, I met a doctor, a club mate of Stoff and Perkeo, a Dr. Schüler. You cannot imagine the joy I felt to be able to speak to someone about shared acquaintances. At first he was not at all pleased that I had gotten into his quarters, but then we became good friends and we are now always very pleased when we see each other. The battles continued and the next evening we were together again. In my room there were injured who were moaning and the next morning one of them was dead.

On the following morning, undetected by the enemy, we left. Our left wing had been victorious, but the less said about our wing the better. For the following day, a Sunday (it was 4 October), a day of rest had been proclaimed. I slept in a warm room and was happy to have such great quarters especially during a day of rest. In the morning I got up, washed myself (I note this here specially because washing oneself was a rarity), and suddenly the news came that the "beaten" enemy had followed us and was near Niedzwietzken (that's what the tiny village was called).

We had barely received the news when the first shell came in, right beside our ordnance park and then, for the first and hopefully the last time, I experienced what flight truly means. Our train of supplies, mixed up with the poor people who were fleeing (because they were afraid of the Russians) stormed down the road in total chaos. It was a shocking sight.

But then something came to light—Prussian discipline! Our men behaved in an exemplary manner. While shells and bullets flew around our heads we all were in our proper positions, directly in front of the enemy. Within 2 hours we pushed the enemy back, suffering great losses, but truly with God's help. If the enemy had attacked less hastily and had waited half an hour, we would now either be dead or imprisoned.

For all this we have to thank our cavalry, which has been operating unbelievably unprofessionally throughout this campaign here in the East, and it is not doing any credit to its name. The enemy did not make the best of their victory (at least that's what I call it). If [the Russians] had sent greater numbers of troops after us, it would have ended very badly for us. But all the better for us! Ever since that day we tell ourselves, "Let's hope things don't turn out as they did in Niedzwietzken."

Let me mention here <u>*Hindenburg*</u>*, a name that impresses the troops more than any other. A short while ago we heard that Hindenburg and his troops were behind the enemy, that he had left Lemberg with his corps to connect with us again (we had mistakenly assumed that Hindenburg and his corps stood near Lemberg). People were almost beside themselves with excitement. You would hear people say, "If Hindenburg were here with us this and that would not have been done that way." Our eastern army adores Hindenburg like no other.*

Well then, on the evening of 4 October we started marching again. The weather gods were kind to us and we were able to ride during a great, not too cold, starry fall night. At midnight we were at our destination at Schareyken, where we stayed until 4:30 a.m.

When we marched through Marggrabowa I was able to send a message to you, that I hope has reached your hands. I had dictated it off my horse to a soldier and handed it to a woman who was to forward it. It was the first time since 27 September that we met with mail service. At 4:30 a.m. we went on to Goldap, where we were loaded onto cattle cars to Eydtkuhnen, of which I have already written to you. Etc.

This great love which unites us both through God's grace
Will through God's grace see to it that we will re-unite[4]

* * * *

By mid-September 1914 the European countries had lost hundreds of thousands of their standing army and ready reserve soldiers and had expended and lost more resources than they thought they would through the entire war. The war was supposed to be quick, decisive, and "over by Christmas." On the Western Front the German invasion of France had come to a halt after surprising resistance from the Belgian Army and a series of massive battles with the French and English. A stalemate line developed from the sea near Ostend to neutral Switzerland.

On the Eastern Front the German Eighth Army had repulsed the invading Russians in East Prussia after the triple battles of Gumbinnen, Tannenberg, and the Masurian Lakes; while in the south a force of more than a million Russians had soundly defeated

[4] In his memoir Riess always replaced any personal text that he had written at the end of a letter with "Etc." However, his wife recorded in her diary that this couplet was at the end of this letter.

the Austrian army at the Battle of Lemberg, and they threatened to take Vienna with a further advance.

The world was shocked as the press reported the vast number of men lost in the first month of the war, and each belligerent country was angered and resolved not to let their enemy escape without revenge. Sir Winston Churchill wrote, "The antagonists, gasping and streaming blood, but still possessed of unmeasured resources, their full wrath unloosed, paused for a moment to rearrange their armies and refill their ranks, to replenish their ammunition and shape their plans anew."[5]

The Germans and Austrians were in a precarious position on the Eastern Front. Both armies were regrouping, refilling their ranks, and resupplying. With luck and skill they were listening to and decoding the Russian command's radio broadcasts, but what they heard was disturbing.

The Russians knew that the Austrians were too weakened by their defeat to attack into Russia, and that the bulk of the German forces were tied up on the Western Front. Therefore they moved three newly arriving armies west, directly against the Germans. They planned for their rebuilt First Army and Second Army to maintain action against the German Eighth Army's front in the north, while their three new armies would sweep in from the south and take the Germans from behind, destroy them, and march to Berlin unopposed with a combined force of five Armies consisting of almost a million soldiers.

The Austrians and Germans together had only about half as many men to stop the Russians. In the north, to counter a Russian attack by the Russian First and Second Armies, the Germans decided to leave a secretly weakened Eighth Army in a defensive

[5] Winston S. Churchill, *The Unknown War: The Eastern Front* (New York: Charles Scribner's Sons, 1931), 232.

position along the East Prussian-Russian border. Riess's unit remained in the Eighth Army for a while as part of the ruse.

Almost half of the German Eighth Army and some recently arrived units from the west were quickly moved south to form a new German Ninth Army. The Ninth Army held the center position of a long defensive line, with the Eighth, at about half strength, to the north and the Austrians to the south. The German generals expected an attack in the center from a force more than three times their number. However, instead of waiting for the Russians to define the battle, the German and Austrian generals decided to attack east before the various Russian corps had finished organizing. With luck they could defuse the Russian onslaught, and with more luck the Germans also might take Warsaw.

Thus the German Ninth Army and all the Austrians attacked on 28 September and in six days of fighting drove the surprised Russians east many kilometers to the Vistula River. Knowing they were outnumbered, the Germans and Austrians prepared for a rapid withdrawal after their advance, with their engineers repairing roads, bridges, and defenses along the way—and setting explosive charges on these same structures to destroy them if that were needed after they withdrew. As the Germans fought at the Vistula River, they were increasingly pressed by the constantly arriving forces of the Russian armies, who threatened to envelope them from either flank. Having defused the Russian attack, the Germans and Austrians quickly withdrew to where they had been. Their rear guard units destroyed all modern means of transportation as they marched west.

With the railroads, roads, and bridges destroyed, the Russians could not support their armies far west of the Vistula River. Therefore they could not advance in strength to attack the "new" German lines. Although Hindenburg had hoped to capture Warsaw, and many people at the time considered the German-Austrian attack a defeat followed by retreat, with hindsight one can see that the operation had been a huge, successful German-Austrian raid that thwarted the Russian plans to invade Germany. It had cost the lives of more than 40,000 soldiers of Germany's new Ninth Army, and tens of thousands more of the Austrians and Russians.

Meanwhile, to the north the remaining men of the German Eighth Army did their best to act as if they were a much larger force. That required constantly moving the most mobile units to reinforce defending or attacking units. 6th Battery was just such a mobile force. Well trained, equipped with very mobile guns, and now veterans of several battles, they fought up and down the front lines, sometimes in artillery duels with Russian gun crews, sometimes in close support of a cavalry attacks, and other times providing accurate fire just over the heads of infantry in defense or offense. Thus Sergeant Riess and his comrades were almost constantly fighting or on the march. Like most soldiers in a war, they did not know, even months later as Riess wrote his memoir, what was happening beyond three or four kilometers from their positions. Nor did they know why they were ordered to move, wait, build defensive positions, attack, or retreat at any particular time.

At the end of September, 6th Battery was part of the right or southern flank of the German Eighth Army, north of the German Ninth Army. As Riess related, they were temporarily shifted south toward Raczki during a terrific rain and hail storm to participate in an attack, probably timed to take some Russian attention away from the Ninth Army's attack against the newly arrived Russian armies. The two-day forced march through the storm cost them a number of men and horses, lost to exhaustion.

When they arrived, 6th Battery deployed in a position supporting infantry just west of the early fighting, near Raczki and Janowka. They held their position, and in two days of heavy fighting they advanced with the infantry and hussars a few kilometers to the east, then fought their way south. There they started taking heavy

Russian artillery fire from the north, east, and south—they were almost surrounded.

6th Battery shifted west in a fighting withdrawal and deployed in and around Marggrabowa. Here Riess sarcastically recorded that the German hussars had been so inept at scouting that the next day, Sunday, 4 October, the Russians were able to launch a combined artillery, infantry, and cavalry surprise attack. The regimental history recorded that in short order the unit had artillery shells, rifle bullets, dragoons (mounted riflemen), and the dreaded Cossack cavalry among them. The men of 6th Battery were in a desperate fight to save themselves and save a large supply convoy, or *train* of wagons, that was just arriving. Because weapons were in short supply, the field artillery men still had not received their carbines and handguns. They could only fight with their cannon and short swords.

A Russian general's report, found after the war, exonerated the German hussars whom Riess scorned in his memoir. Evidently before the Russians arrived, the German hussars had made a scouting sweep of the forest and fields to the east of the German lines and found nothing to report. After the hussars returned to the German lines, Russian scouts spotted from a distance that a German supply convoy was moving south, seemingly only lightly protected. The Russians decided to attack it quickly in case there was a strong German force nearby. They sent every mounted regiment in the region for the attack: the Czar's uhlans, hussars, Cossacks, dragoons , and horse artillery batteries.

The mounted Russian regiments arrived and rode through the forest, the artillery unlimbered within the western edge of forest, and the dragoons dismounted. The effect from Riess's viewpoint was that without warning artillery shells, infantry, and hundreds of charging cavalry burst from the forest.

The men of 6th Battery, taken by surprise, made a fighting retreat to a better defensive position held by the regiment's assigned jägers and turned against their foe. They fired their cannon for almost two hours against the attacking cavalry and infantry and then, after beating back the attackers, 6th Battery became locked in an artillery duel with the Russian horse artillery. After an intense shooting battle the Russians withdrew. Their mounted forces had made only one charge and withdraw because they had planned the event to be a raid—attack quickly and then withdraw.

No one recorded why, but that evening, after the fighting had stopped, Captain Kujath designated Riess a squad leader, in command of one of the four gun crews in 6th Battery. Possibly he was replacing another sergeant who had been shifted up the command structure or had been a casualty in the battle. He did not record his feelings about that.

Having fewer than half as many men as the enemy facing them, the German Eighth Army's command had to keep up appearances with action. They correctly calculated that the front in this area was temporarily stable, so they ordered 6th Battery to march north, where fighting was intensifying. On the previous day the Russians had launched a serious attack near Eydtkuhnen, held by a German infantry division. 6th Battery force marched to Goldap, where they boarded a train to Eydtkuhnen. They settled for the night west of that important rail crossroad, as a major battle continued through the night on the east side of the town. In the morning they took position on the division's right flank and, because the Russian attack had been thwarted, they had a day of full-alert "rest" and, as Riess wrote, a night of "general drunkenness."

In the morning the Germans counter-attacked, beginning a two-day battle around the villages just east of Eydtkuhnen. Riess, newly in command of a gun crew, was fighting in a confused series of short battles in various places, sometimes firing in close support of attacking infantry and at other times firing from a distance of 3 kilometers at moving enemy troops, defensive trenches, or artillery. At the end of the second day the Russians withdrew to the east, and 6th Battery quartered for a day near Wladyslaw, a village outside Eydtkuhnen. Here, on 10 October, Riess wrote to his wife in Berlin, sending the above letter and couplet that appear in his memoir. His letter arrived three weeks later, just before her twenty-third birthday.

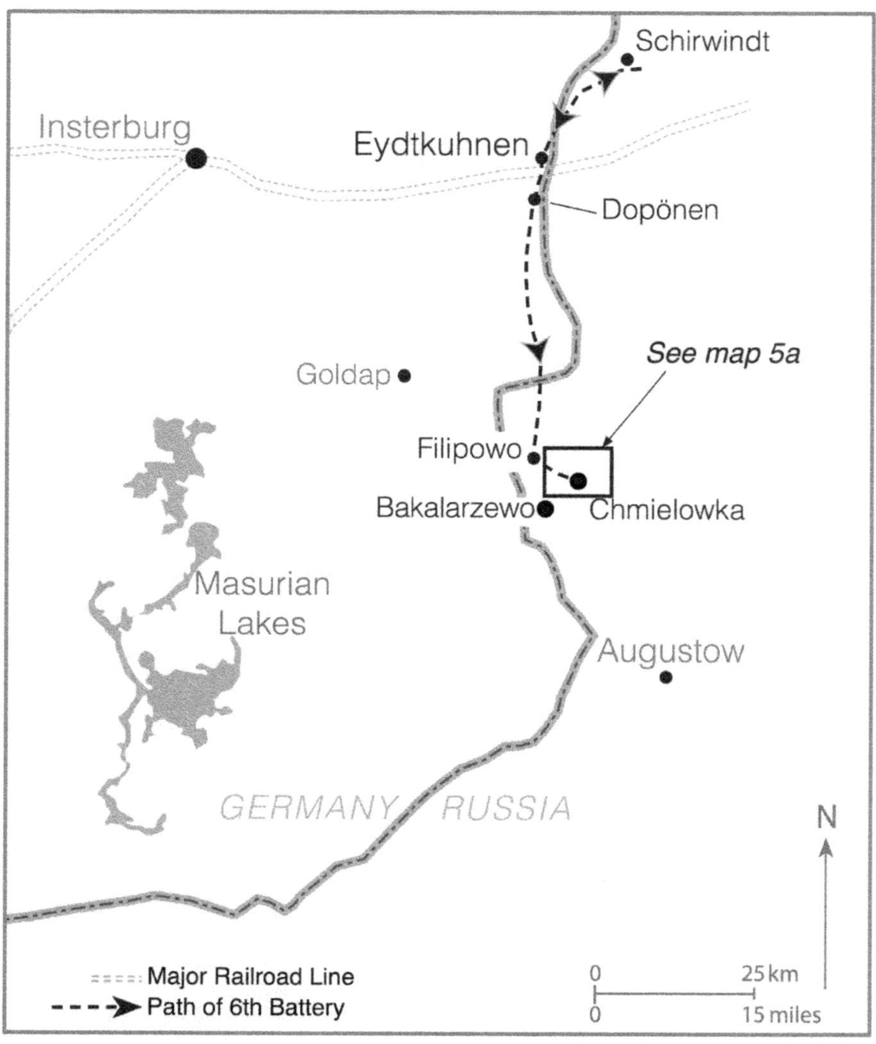

Map 5. 11 October–1 November 1914. Eydtkuhnen to Schirwindt to Chmielowka.

CHAPTER 5

South, Along the Russian Border

11 October–1 November 1914

We did not have much time to enjoy our rest. Already on the 11th, the enemy marched toward Schirwindt [Kutuzovo] again, and since we had done so well in the past, we were to go again. As early as Barzkehmen [Bartztal] we met with [artillery] fire, a sign of how close at our feet the enemy was. In that very close proximity to the enemy, near Barzkehmen we spent the night. At 1:30 a.m. a Russian battery was stormed. It was a horrible picture to see that stormed battery in the morning of the 12th. Horses and men were lying in mingled heaps among the guns. A story made the rounds about a Russian lieutenant who, having been urged to give himself up, answered by slapping our German officer in the face; and after his wrists had been tied he broke free and did the same once again. He knew what the next few minutes would bring. In my opinion this was a sign of immense heroism, something the Russians rarely showed in this campaign.

I mentioned it previously and want to point out again how desensitized we had all become. When we stopped beside the corpses of men and horses, we had breakfast right there and then.

After that we went into position. Five gun emplacements in three hours—we were rushed as never before. [We made such progress that] I thought we were supposed to celebrate the evening in St. Petersburg. Altogether, we captured 30 guns, 5 machine guns, 3,000 soldiers; and many, many dead were lying on the battleground.

The weather had been good while we were within Germany. That night we went across the

Russian border again near Schirwindt, and immediately the rain started so that we sank up to our knees into the muddy soil. Out of the rain I moved into a house with my men. The inhabitants of the house had lost their daughter to a bullet that day and they were inconsolable.

Despite the rain we had to return to Germany in the evening. Again, we had fulfilled brilliantly what we had set out to do near Schirwindt and returned to our old drafty barns, where we were miserably cold after some downpours. The next morning we were able to go inside a house to dry out and then we went to Barzkehmen into really nice quarters. We were back in Germany and immediately the sky cleared up.

On the 14th we were back at Eydtkuhnen and stayed there until 18 October. It was a great time! On the 16th I rode to Insterburg and was able to telephone home. I still feel emotional thinking about it today. Not having spoken with Gertrud from 3 August until 16 October and then suddenly hearing that voice, it was wonderful! All my memories of Insterburg are wonderful: when we marched in, when I was able to make a telephone call, and one more time that I will describe later. With gratitude I remember the ladies of the telephone station in Insterburg who, when I told them that I had not spoken with my wife in months, let me make that call despite the fact that this was strictly prohibited. I called on these kind ladies another time later, with the same success.

Alarm was given again on 18 October. We had to move forward on our right wing as a Russian breakthrough was imminent. We bivouacked at night and, it being 18 October, it was very cold. We didn't have tents any more. On the 19th we awaited the enemy. But we saw nothing, only the mountains from which the enemy was supposed to descend. We waited in good spirits because we were located in very secure positions for the moment, from which we could send over our first salvos.

The enemy didn't show up, and around midday we marched into Dopönen [Pokryschkino], which had been completely shot to pieces. We moved into the quarters of a wealthy gentleman farmer, who lived in a basement—which was all he had left. The next morning we were supposed to move on. Our guns were stationed at a distance of about 15 minutes from the place where I had taken quarters with my section, and when I was riding over there with my men the next morning, I got lost and appeared 1½ hours late. I didn't feel good about it but I got off fairly well. I was received without any harsh words and a short time later we all went back to our old quarters, as the enemy still was not moving. And we never did meet the enemy at this spot. Later on, when the Russians finally came over the mountain, the troops who had taken our position gave them a proper welcome. Who does the job is not that important, as long as the job gets done!

We stayed in Dopönen on alert until 26 October and lived a great life there. Apple pies were baked for us, there was freshly slaughtered meat every day, and the farmer's wife was an excellent cook. We had to sleep in the kitchen on straw covered with bedding, but it was a good time anyway.

On the 26th we left and went from Dopönen to Budweitschen. We were in bivouac there and the next day went off to Bartnagora. We were approaching Augustowo again, where we had never been lucky. On the 28th we went into position near Chmielowka and by afternoon we had to change position already.

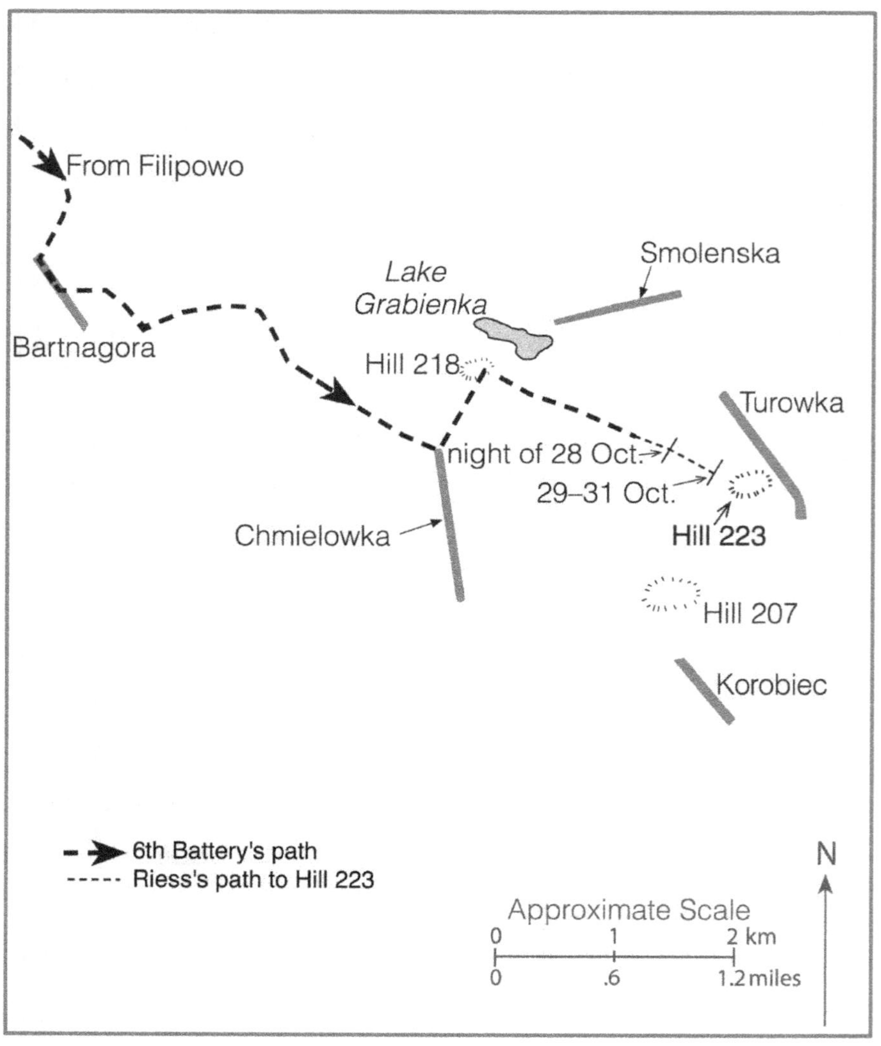

Map 5a. 27–31 October 1914: The battle for Hill 223. The hamlets were communities built along local roads.

During that evening I was sent within 200 meters of the enemy with a platoon [two guns and their crew] in order to assist our infantry in an attack expected any time. I can only say that I did not have a good feeling about this. If the Russians had attacked and our infantry had had to withdraw, what would have happened to my guns?! But all went well and next morning I was back with my battery safe and sound.

The Russians had determined our position between Smolenska and Turowka in great detail and opened such strong fire at us that we were ordered to leave all our guns behind and flee into dugouts that the infantry had prepared beforehand. Heavy enemy artillery fired at us with such intensity as never before seen. It is a miracle that only our observer's wagon (on which our captain had stood a minute earlier) and our 3rd ammunition carrier (behind which I had sat just beforehand) were destroyed. Miraculously, no lives were lost. Afterward we had to use ropes to pull each gun out of position because if we tried to get very close to the guns ourselves, the Russians shot at us immediately.

Lieutenant Deuss was sent as an observation officer to the most forward trench that evening, where he participated in the storming of Hill 223. It was to be his last. We set up a memorial in his honor under a fir tree. I don't want to give great detail about this attack on Hill 223 and want to skip what happened on 30 and 31 October and 1 & 2 November, because these heaviest battles that I encountered are described well in my letter [to Gertrud]:

> *Much has happened since you have last received detailed news from me. Many hard battles we have endured with all the associated hardship, battles that were heavier than ever and put all the previous ones into a different light. With God's help I have come through it all, but I have to confess that I have never been under similar fire before. In this battle for the first time I had the feeling that something could really happen to me. At other times I usually became thick-skinned when bullets flew by and therefore I had thought that nothing could disrupt my peace of mind in combat. But I have been in situations here, where I doubted very much that I would get out unscathed.*

Recalling all that difficult time is quite all right, because it is all over and done with now. To tell you the truth though, it makes you feel cold all over when just 5 meters away from you one of those big things hits or when the trench in which you are located is being taken under attack by enemy fire. All this I will describe in greater detail later.

It was very important that on the day I left my really nice alarm quarters [forward resting quarters for "ready" units] in Dopönen, I received your fur. It is wonderful, it fits well and without it I would have been terribly cold, as the temperatures went as low as -4°C [25°F] and several times we spent the entire night outside.

Other times we spent the night in holes we had dug in the ground and which also, for obvious reasons, are not very warm places. Even when, to keep warmer, we slept so close together that we couldn't move, even when we used blankets and coats, we were still quite cold. I want to tell you something strange: if I sleep outside, even when it's really cold and even if I lie on wet things, I never seem to get a cold. However, as soon as I am in a warm room for a couple of days a cold arrives.

Well, as I said, the fur and the blankets were very, very pleasant and undoubtedly they kept me from having to leave due to a bad cold, as some of us had already had to do. What the human body can endure! Just think, at -4°C you are outdoors day and night! One can get used to just about anything.

But now I will begin to tell you how things were, what happened, and how we felt when we had to withdraw, undetected by the enemy, in the middle of the night. There was, as official reports noted, "nothing new along the eastern theater of war." Near Smolenska and Turowka close to Suwalki we received a "sound thrashing," as my friend Schmidt calls it. Where he is today nobody knows. I hope he is in Russian imprisonment; he became one of the first victims at Prassnitz.

We were outnumbered ten to one, and we withdrew in order not to suffer greater losses. The enemy had taken extremely strong positions and storming against these was sheer madness that we paid for very, very dearly and bloodily.

So, on 26 October we left our alarm quarters, where we had led a very decent existence. We had to live in a basement and sleep on the floor, but otherwise life was very comfortable there. The food was excellent and we all recovered nicely. Generally speaking, food and rations in our battery are very good and, with the quantities I eat, in part from our battery rations, in greater part from my own supplies, I don't think I have lost any weight (I was wrong there, I lost 12 kilo [26 pounds] altogether).

We knew we had a long march ahead of us. After a very short, very unpleasant night's rest with a German family, who do not deserve to be called that (we found quite frequently that Russian inhabitants treated us better than the German ones, who were only afraid that they would lose something) we went across the border again.

The mild weather we had had in Dopönen disappeared and it became quite cold. Via Filipowo the road continued, except that one could no longer call it a road. We could detect very few traces of it, only the usual Russian mud and the horses had to wade knee-deep in it at times. The distant rumbling of guns that started coming closer reminded us that things were getting serious again.

We were given one more peaceful night and the next morning we went into position close to the totally destroyed Chmielowka. We received heavy gunfire as soon as we started moving into position and our 4th battery had its first dead man during approach and even before dismount.

Our battery had been given a bad position and by afternoon we had already had a change in position, which could have ended tragically for us, as it turned out the next day. At first, an order came that after darkness one platoon (2 guns) was to go forward to our infantry, within 200 meters of the enemy. Our captain entrusted me with this honorable yet questionable job. I had never led a platoon, was to find a position all by myself, was supposed to support our infantry, which was expecting an enemy attack, and was to bring my guns and men back home intact and well. I prayed, "Dear God, let this all come to a good end!" As drastic as it sounds, I was scared shitless!

The Russians are great swine, but about this campaign there is one good thing I can say about them: they left us alone that night. When I think back to that night, I thank them for it even now.

At dawn we moved away and were happily back with our battery an hour later. We thought that after having had to be on watch all night, we would get a bit of rest, when suddenly heavy enemy artillery started shooting at us with such strength as never before. Every shot landed in our battery and our captain gave orders for us to leave the guns and move to dugouts that had been set up by the infantry in previous fights.

What it means for a Prussian artilleryman to leave his gun behind, everybody, even the non-soldier, can imagine. It was terrible! Our observer's wagon, on which our captain had stood not 5 minutes before, looked like a heap of debris. The 3rd ammunition caisson, behind which I had stood, was pierced everywhere. And if these dogs could have aimed better, the entire 6th Battery would have become a jigsaw puzzle within a short time.

After an hour we dared come out of our dugouts. But immediately the enemy opened fire again. Back into the dugouts we went. Much later we moved our guns and ammunition carriers out of that position one by one and miraculously, thank God, we were able to do so without any losses. Another position received our intact carriers and guns.

I can tell you, I don't want to experience hours like these often; my nerves were wearing thin.

(It was at that point that I had my first real gallbladder attack. For understandable reasons, I didn't mention this in my letter.)

Unfortunately my pen isn't eloquent enough to describe all the incredible dangers our battery faced. When a day starts badly it ends badly too! That same evening an officer was to be sent forward as an observer, something I will explain in more detail later. Our Lieutenant Deuss was ordered to do so. During heavy combat he was hit by infantry fire and lost his life. We buried his body under a beautiful fir tree.

He died during an assault that he was leading against Hill 223 near Suwalki (an unforgettable location for me), which was taken by us after half an hour of fighting. Our 61st Reserve Regiment lost 900 men, literally 900 men within that one half-hour. And what was accomplished overall? At night the order came for us to clear from that hill that we had conquered by way of such heavy losses. It was an order that has puzzled all of us since then, because all losses we suffered in the days to come, and our inglorious withdrawal, were a direct result of clearing from that hill.

A great many who were wounded still lie up there. Those who had not been transported away in time, died of hunger and in pain, or froze to death. We had this terrible awareness that was with us in the days after—these poor people are lying up there, while they are awaited at home. What a misery war is—and after all is said and done, we just proceed to the order of the day!

On 30 October I wrote in my diary: "Heavy fighting continues. We are constructing a cave for ourselves." This was the hole I mentioned to you before. The men build these caves with great skill, directly beside a gun. Each gun has its own cave and now the men compete with each other for constructing the best cave. Each of these is about 1 meter deep and 4.5 meters wide, at the top it is covered with boards, earth, leaves, and straw and the interior walls and the floor are also covered with straw. When the men get cold during the day, they work on their "homes." And in the evening, when the lights shine inside and 7 men lie in it, it can be comfortably warm, especially when they also get heated from the inside with "gasoline," as rum and brandy are called.

One of our officers was sent to the rifle trench [at the front line] for observation. At that position it is easy to check on all the shots that are fired by the battery because, for the most part, the target can be seen very clearly. With the aid of a telephone, the wires of which are laid down at night, the battery gets told how to shoot and in which way and from which side our infantry is getting attacked. After 24 hours the lieutenant was to be relieved. "Cavalry Sergeant Riess, you will relieve Lieutenant Glaubke!"

At dusk I get on my way, receive relatively mild fire, yet see all the damage these big [artillery shells] have caused—big holes, a meter deep. Houses that were here two hours ago have almost completely disappeared. There a rifle, speckled with blood, that its poor owner had to drop. There a field pack with all kinds of things. It belonged to an injured man for sure; one who was unable to drag this heavy burden to the dressing station. Over and over one becomes aware what a terrible disaster war is!

So I take off with my telephone operator and we reach the trench after half an hour. The lieutenant helps me orient myself, easily done in the darkness: not more than 50 meters in front of me is this awful Hill 223 with all its horror. Yet I spent a rather pleasant evening with the officers of the machine gun division who were in that trench and had built themselves a great cave. Many lights and even more men together, along with some blessed brandy, delivered the necessary warmth. Since the enemy left us alone, it was all very, very pleasant.

I can tell you about a funny little episode. Naturally, in the very same cave where the captain and first lieutenant reside, also were the orderlies, in this case older reserve men, genuine natives of Berlin. The captain's orderly, a Mr. Wilhelm Meier of Pankow, is quite a character. Conversing with the officers I told them who I am and what I do.

When this Wilhelm Meier of Pankow hears that I am of G.G. [Gebrüder Grumach—a large, well-known department store in Berlin at that time] he starts howling and says, "Well Captain, that's where I buy everything. All of us do. A really great place!" The captain says to him, "Meier, you are only schmoozing. You just want that cavalry sergeant to give you a cigar!" But then the other orderly starts singing the praises of G.G. too, and the captain is won over, and we all have a good laugh, something we wouldn't ordinarily do in such a situation. G.G., a famous name even in the trench between Smolenska and Turowka!

At 1 o'clock in the morning the regiment was relieved by a battalion. One regiment with 3 batteries had already left us earlier and successfully confronted the enemy near Schittkehmen. This regiment had taken along 2 batteries and only one remained to fill the same place that had been filled by 2, just 2 days ago [meaning that five batteries had departed, leaving only 6th Battery].

When the troops left, the trench in which I was stationed was supposed to be filled anew. When nobody showed up, and when at 3:30 a.m. the telephone connection was disrupted, I had to move away before it became light, and returned to the battery. I reported in and was going to rest, since I hadn't slept in 36 hours. Instead I had to go to the captain, who told me that I had to try to get back to that trench and urged me, in God's name, not to put myself in danger. Those were very nice words, easier said than done, impossible to follow.

But I did want to get to that trench, under any circumstances. For a great part of the way I was able to cover my steps, even though shells would always hit beside the telephone operator and me. When we were approximately 300 meters from the enemy, we hid behind a house and waited until things calmed down. Then up, march! Down! Then up again, down again. Beside and above us shells exploded, and when we happily got to the trench, the trench was getting fired at.

Backed up almost all the way against the wall I observed our shots and the enemy shots. As already mentioned, the ditch was supposed to have been occupied by infantry again early that morning, but nobody had arrived, and the entire flank was covered just by a cavalry sergeant and one man. If the Russians had known that! I reported to the captain with exactly these words, "If a flank attack will occur, I have only one weapon, the telephone, and that's what we will throw." Apparently these words were cause for great amusement.

I was ordered to stay in that trench the whole night. At 3 p.m. an order arrived that we were to break up telephone connection at dusk and return. I was so happy I was beside myself. The previous troops had taken everything out of the cave, it was freezing cold, and I had to stand outside to observe the entire time. As soon as it became dark, we started back, with clattering teeth, having eaten nothing for 24 hours.

Upon returning to the battery I received great praise. When I went to the captain, all the officers were there. The captain said to me, "I give my full recognition for your excellent and brave conduct. You have done very well." Then I had to tell about my expedition to the trench and when I had finished, the captain told me that we were going to withdraw to Germany under the cover of darkness. Mixed feelings on my part! On the one hand, Germany, a very tempting proposition; on the other hand retreat, a depressing thought. But I was very glad that I had gotten out of this situation unscathed.

We marched all night until 7 a.m. Marching through villages, the complaints of the inhabitants were always the same: "Now the Russians will come back." And they were right. The enemy, not having noticed our withdrawal, fired mightily at the position we had already left. I hope they wasted plenty of ammunition. At least these rounds didn't hurt anybody. Yet the feelings came back about all those poor souls up there on Hill 223, who had fought for the fatherland as we did and who shabbily had been left behind.

That night, because of fatigue, I twice almost fell from my horse. Seventy hours and not a minute of rest, 20 of which I had to observe steadily and precisely. I was quite done. During the day I couldn't find any rest because I was over-tired, but at night I slept like a log. Again, we were out of Russia, a place we detested very much. Etc.

* * * *

Throughout October 1914, Sergeant Riess's unit remained with the Eighth Army in East Prussia. As he admitted, we can see that he was becoming desensitized to the battles. He left out of his memoir many smaller battles in which he was involved as 6th Battery moved toward its various objectives. It is not clear if he was too busy and weary to write about them in his field diary, he had forgotten about them, or he consciously decided to skip over them. However, the official history of the 36th Reserve Field Artillery Regiment provides many details and maps.

On Sunday, 11 October, the men of 6th Battery were hoping for a day of maintenance and rest, but at 11 a.m. they were ordered to mount—at least one Russian division was attacking in force north of Eydtkuhnen. The field artillery was to reinforce the northern flank of the infantry at Schirwindt, but well before they reached Schirwindt they were in the battle. While the infantry division and some of the artillery held the line, 4th, 5th, and 6th Batteries advanced and silenced two Russian artillery batteries in the late afternoon.

Before dawn the next day, 6th Battery advanced in support of an infantry regiment as far as the previous Russian positions, ate breakfast there among the dead men and horses, then rapidly pursued the retiring enemy. The Russians made a fighting withdrawal, with their rear guard and field artillery repeatedly establishing defensive positions, forcing the pursuing Germans to mirror their moves in order to fight and advance.

6th Battery, supporting the infantry, fought their way to Schirwindt, where the Russians had destroyed the bridge across the river, sacrificing the men of their rear guard on the German side of the river to be killed or captured by the pursuing Germans. After the German infantry captured hundreds of marooned Russians and their equipment, the field artillery forded through deep water onto Russian soil to support another infantry regiment already across the river. The men and horses were exhausted, and could hardly advance with their guns and wagons through the mud on the Russian side, but then they were ordered to withdraw back through the mud and across the river before dawn. They were needed elsewhere.

The predawn hours found them again in Germany, beyond exhaustion, marching south toward their previous quarters near Eydtkuhnen. However, by 6 a.m. they received orders to pursue retreating Russians just north of Eydtkuhnen. They did not catch the enemy, and after collecting equipment and supplies the Russians had left, they turned once again to their quarters west of the town. They had collected Russian field artillery guns, machine guns, and many supply wagons and caissons, but the men especially valued good boots. The two-day battle for Schirwindt was over, leaving towns and hamlets destroyed, hundreds of dead, thousands wounded, and 4,000 Russian soldiers taken prisoner.

By mid-morning of 14 October 6th Battery had settled back into their quarters around Eydtkuhnen. That day a train arrived at the station with supplies and Red Cross gifts of port wine, preserves, and other items for the troops. The regiment was able to stay there for three days to repair, resupply, integrate new horses, and rest. This was when Riess rode a few kilometers west to Insterburg and was able to make a telephone call to his wife in Berlin. He must also have ordered flowers to be delivered, because on her birthday at their Berlin home she received a bouquet of carnations with a message from him. She pressed one in her diary (see back cover).

On 17 October intelligence officers reported to the Eighth Army headquarters that the Russians were preparing major artillery and infantry positions south of Eydtkuhnen, therefore 6th Battery moved south and established new positions near Dopönen. There they dug in and waited for the Russians to attack, but nothing happened, and so they remained in their positions through 25 October. The men found quarters on nearby farms and in hamlets.

While they remained at the border on alert, wranglers brought another herd of horses to replace fallen animals and provide a few reserves for the future. Properly sorting and training the horses for their new tasks kept many of the men busy for the week.

On 27 October the battery marched south in a cold rain to support the infantry and cavalry at Chmielowka, a few kilometers west of Suwalki, where the Germans thought the Russians were massing for an attack. The German generals decided to attack first, against well-fortified enemy positions, and therefore shifted a number of large artillery units to support the advance. They included many field and heavy cannon and howitzer batteries. The night turned cold, below freezing, allowing the horses to move the artillery batteries without much difficulty in the early morning. However, men and beasts suffered each night as they bivouacked where they could without shelters, not even having tents any more.

On the night of 27 October, 6th Battery supported the north flank of the German front at Chmielowka. The battery found already dug trenches on Hill 218 and bivouacked there for the night; they had the southern shore of Lake Grabienka on their left (north) and the enemy in front, on Hill 223, to the east.

The Battle for Chmielowka began the next morning, in a dense fog, with a salvo from the German heavy artillery that was quickly answered by the Russians. A long distance heavy artillery duel continued through the morning and then the Germans began hitting the Russian trenches with heavy howitzers. The men of 6th Battery could only watch in the morning as their small cannon did not have the range of the heavy guns. They moved south off their hill, to reinforce the infantry line near Chmielowka, but when they arrived there the position was not good, so they shifted north again and dug in between Hills 218 and 223.

Though seriously outnumbered, the German infantry prepared to attack the next morning and needed a solid front line in case the Russians decided to make a night attack first. Therefore 6th Battery spent the night in their assigned position just behind the infantry, while receiving heavy but inaccurate fire from the enemy artillery. They had orders to fire only when necessary as there was a shortage of ammunition for their 77 mm guns. This was the night when after dark, Riess was sent forward in command of a platoon (two guns), who advanced quietly to within 200 meters of the Russian infantry to provide very close support for an "early warning" defensive infantry line. Luckily the Russians did not attack, so Riess and his platoon returned to their previous dugout position just before dawn.

6th Battery was positioned far forward of the other field artillery, to support the infantry with close-range fire against the Russian field artillery. Thus with dawn they were targeted by Russian heavy artillery for several hours, and infantry rifle fire hit their equipment and supplies, but none of the men in their shallow trenches were hit. Between the artillery barrages that day, during breaks possibly caused by the Russians' slower resupply system, the 6th Battery men worked by hand with ropes to pull their guns from west of their position back to their limbers. The horse teams then moved them out of range, to the western end of Lake Jaciewo. The German generals postponed their attack for a day while the heavy artillery and howitzers pounded the Russian troops' positions.

The next day, 6th Battery advanced on the enemy, providing close support for the infantry attacking the well-defended Hill 223. The rapid, low-trajectory, accurate fire of the battery's four guns was needed to destroy the Russian fighting emplacements as the infantry stormed forward and took the hill, suffering very heavy losses. After seeing the array of forces at the end of the day, the division commander felt Hill 223 was too far forward of the battle line to defend. He ordered everyone on Hill 223 to withdraw. The 2nd Battalion withdrew west, out of range of the Russian heavy artillery, except for 6th Battery, which remained forward with the infantry just west of Hill 223. This was when Riess spent a quiet night as a forward artillery observer in the trench with the machine gunners, drinking brandy and smoking a cigar.

The next morning, on his wife's twenty-third birthday, his telephone line back to headquarters was broken, possibly by artillery fire. As soon as he returned to his battery he was sent back to the forward trench, accompanied only by a telephone operator, who evidently laid a new phone line. When they reached the trench, they found themselves alone, as the machine gunners they thought were holding the trench had left.

Riess remained there as the battery's forward observer, exposed between the two armies through the day's intense artillery duel, calling in directions to the artillery. He and the phone operator were less than 300 meters downhill from the Russian lines in broad daylight. Yet because the regiment, already three months into the war, still had not been issued personal weapons, neither of them had a firearm to repel even one enemy scout. Overhead artillery shells screamed, as 6th Battery and a Russian field artillery battery began a duel that day, and, probably with the help of Riess's observations, the Germans were able to destroy the Russian guns and some of their men.

Riess did not know that during the day his whole division was being withdrawn, with the four guns of 6th Battery firing away at Hill 223 to cover their retreat. Thus Riess and the telephone operator were rear-most of the rear guard, and they were unarmed. The Russians must have thought the machine gunners or infantry were still holding the trench, as they did not try to take the trench until after Riess had left it.

As soon as it was dark, Riess and the telephone operator returned to their battery, and 6th Battery with the rear guard infantry withdrew to Germany. Though "low on ammunition," during the battle each gun crew in the battery fired more than 200 rounds of ammunition. When writing his memoir months later, Riess still thought he had been part of a general advance that had failed. However, the German generals' plan from the beginning had been to attack and withdraw. The German Eighth Army's preemptive strike to defuse a Russian attack, including the grim losses at Hill 223, had been costly but effective. Meanwhile, a major crisis was developing to their south.

Map 6. 1–30 November 1914. The Battle for Lodz.

CHAPTER 6

Into Poland with the Ninth Army

1-30 November 1914

Early that morning we left Gurren. German roads and German weather! It felt like being at a field exercise. That day when I came down from Chmielowka, I had never felt that we were retreating; instead I had felt predominantly that I had been lucky to have gotten out of the terrible situation up there. I had gotten out of a situation that, as my captain told me a few days later, earned me the Iron Cross.

From Gurren we went to Gross-Wischteken, and on the way there the captain ordered me to go to Insterburg where I was to spend the night. I think I never grinned as broadly as I did then. I was going to be able to use the phone, sleep in a bed, bathe—the latter for the first time during the campaign, and, even more important than all other things, speak with my family. I can only repeat: Blessed Insterburg! The ladies at the telephone station, having helped so very kindly in the past, helped me again, and long, detailed conversations with family in Berlin and Stettin made me a very happy person for several hours.

All kinds of rumors spread that we were going to be entrained. Nobody knew where to. Different destinations came up. Some said to the west, others said to Turkey, others said to garrison some fortress, some others to Warsaw. The last ones were correct. We didn't actually get transported to Warsaw, because our troops located around Warsaw had been forced to retreat in forced marches thanks to *our dear confederates* [the Austrians] who had failed once more.

For that reason our troops had lost connection to the Austrians and a catastrophe could easily have been the consequence. However, our brilliant leader Hindenburg had worked out the withdrawal in such a clever way that the end result was a great victory for us. We haven't returned yet into the same positions where our armies once stood, but nevertheless we have done very well in capturing an incredible amount of supplies, materiel, and prisoners.

I spent a great evening in Insterburg. To top things off, I even went to the moving pictures, but I didn't last more than 5 minutes.

Next morning we were entrained. We still didn't know where we were headed. Suddenly we stopped at Thorn. Our guns and horses were unloaded and on we went to Klein-Morin. What incredible quarters we had there! Our dear hosts, Mr. Hammermeister and his wife, knew how to please us. We stayed there until the 10th, one day of which I spent with our good friends the Frombergs in Thorn, who told me how my dear ones at home were doing. It was so nice to be with people who take a real personal interest. It felt very, very good and with a light heart I went back to my battery.

We had really been spoiled those 5 days there. On 10 November, however, when we went from Klein-Morin to Serovski we felt a marked difference. We spent the night in a foul-smelling shack and now the most exhausting time period began. Starting on 10 November up to the day of my departure we didn't have one day of rest; and even after I left it took a while longer until my battery was allowed just one free day. On the 11th we came upon the enemy and were met with strong resistance. On the 12th too there were battles at various locations, but we continued to move forward.

Fighting continued on the 13th and the enemy, seemingly not very strong at this location, hastily withdrew. That evening we took quarters at Wieslawice, a beautiful castle with stables and beautiful horse equipment the likes of which I had never seen. I think the owner must have wondered the next day how much worse his horses had become overnight. Some of our dear animals, which had been in the campaign from the start, were left to rest at Wieslawice. But since we had to have a full number of horses, they were "exchanged."

The estate looked beautiful from the outside and showed off the owner's wealth on the inside, yet half of our battery caught lice on this property and its surrounding farms. These critters must not like me much. I never had any lice the entire campaign, primarily thanks to two people: first, my wife who constantly supplied me with clean clothes and sent quantities of insect repellent, and my dear old Sell, who made sure that I actually put on these freshly washed clothes. Most of us didn't even think of changing on our own.

From Wieslawice via Kowal we went to Gostynin, where 3 Russian corps were located. Heavy fighting was expected for the 14th and started for us with a bivouac without tents on 14 November. If you were to walk around Berlin on 14 November without a coat, you would be pronounced crazy. Yet where we were it didn't prevent us, or most of us, from sleeping relatively well, for the most part without blankets. The 15th of November was a very successful day for us. Fighting was heavy, and the enemy must have lost many men, as did we!

Into our hands during these days fell 6,000 prisoners, 6 guns, including 2 heavy ones, and 15 machine guns.

One has to realize that our 1 corps went against 3 Russian corps and they [the Russians] got so wiped out that they had to withdraw hastily. We followed the enemy via Bialtotar, where we saw a terrible battlefield. We arrived in Konstantinowo around 1 a.m. in freezing cold weather and we bivouacked at no less than –7°C. When I woke up the next morning, I thought I wouldn't be able to move any limbs because they would be frozen. But things worked. Things work when they have to. Just half an hour later we sat on our horses again, without anything warm in our stomachs.

Oh God, what a sight this battlefield was that we were riding by! I remember a Russian battery positioned at the edge of some woods had been shot at by a heavy battery from our side. From one tree an arm would be dangling, from another a leg, here a torso, and between the demolished guns lay the horses. Even for us, who were used to cruel sights, this was horrible.

We pursued the enemy going through Gostynin on the 16th and soon after Gostynin we ran into the enemy's rear guard and they gave us strong resistance. At night we had billets in Gatschno. On the 17th we met up with the enemy near Gudinowia. Actually, we were still in marching columns and not yet prepared for attack when heavy infantry fire came our way. We stayed there that day and that night. The next morning the [Russian] rear guard—that is, of course, all we dealt with all along—had retreated again and we went from Gombin to Kamin. We were to stay in Kamin that night and then go into firing position the next morning.

Suddenly at night we were woken up, as fire had broken out in one of the houses that we occupied, surely set by a Russian. It was a horrible sight! A poor fellow of ours lost his life. Terribly burned, he came out of the fire screaming, and shortly afterward he died in terrible pain.

The next morning we went into firing position. We were told that the enemy, who was mightily strong, was completely surrounded and now the big "Sedan" day was approaching.[6] So we waited at Kamin, waited for the poor enemy, somewhere, maybe along our own front, [the enemy] would try to escape its "inescapable" fate. But where were they? At the last moment the enemy had found a way to slip through and get out of the frying pan once more.

So on we went, marching! We marched until evening through pretty countryside, on bad roads, passing some beautiful estates, and planning to take quarters that evening. We received strong infantry fire from the village that lay ahead of our new quarters and stayed in combat for about 2 hours. Then we took quarters at the beautiful estate Luszyn.

The next morning, close to Luszyn, we went into position. I was behind the 5th ammunition carrier when I noticed that my 4th gun had turned toward the wrong direction. Despite the warnings of the chief of section not to appear from behind the ammunition carrier, because the enemy was well aware of our position, I went over to the 4th gun. That very minute the gun crew who stood behind the

[6] Probably a reference to the 1870 Battle of Sedan, a great victory for the Prussian Army.

5th ammunition carrier was hit. One man's head blown off, the other unrecognizable—how can one but believe in fate!

At night we went into quarters in Stempow. We left Stempow and, after going past Dlugi, we went into firing position. From there we went into a new one past Slawkow. Near Dlugi we saw many burned corpses and could be quite certain that a first aid station had earlier been hit by indirect fire from our artillery.

I mention this on purpose because I would like to make a point here. We often read in our newspapers that enemy indirect fire hits our first aid stations. For the artillery it is impossible to know exactly what they are shooting at.

Earlier on we had received orders to take Dlugi under fire and we had done so. Had we been able to see at a distance of about 5,000 meters what we were shooting at? It is quite possible for a first aid station to get hit inadvertently; and I do assume that our enemies, as well as we ourselves, do not intentionally shoot at such targets.

We went into quarters near Slawkow. The strongly superior enemy retreated and we continued to stay there on the 22nd as our main objective had been to hold the line in Slawkow. On the 23rd we stayed in our position until evening and then moved from Slawkow to Sobota. On the estate Sobota we spent the night in sheep stables, the entire battery, men and horses. Some of us received souvenirs [lice or fleas] from the sheep, others from the horses. The next day we went from Sobota to Monkolice. Again we were met with very strong enemy fire; we lost 2 men and took quarters in Monkolice.

On the 25th we continued to Glowno via Soppel. Combat here and there, and when we arrived in Glowno that evening we hoped once more, but to no avail, that we would be allowed some extra rest. Nice quarters in Glowno—a bordello with two more or less grimy "beauties," I noted in my diary. These two young ladies were dreadfully surprised that we didn't make use of their services.

Originally our supreme command had planned on a march directly from Gostynin to Lowicz. Then news arrived that our 25th

Reserve Corps was completely surrounded by the enemy and our regiment as well as a cavalry division were sent to Glowno to come to the assistance of the 25th Corps. When we arrived there, this unlucky 25th Corps, not having done well in the war, had not only gotten itself out of the situation on its own but had also taken a great number of prisoners and captured many, many guns and other war materiel. As I just mentioned, our regiment had been sent to help the 25th Corps and a [69th infantry] brigade in our division had received orders to take Lowicz that evening. Considering that exactly 3 weeks later it would take 4 corps to take Lowicz, and comparing this to the order given on the 25th, one can only conclude that unfeasible demands are made of our troops at times.

If ever an unfeasible order was given it was to ask 69th Brigade to take Lowicz that evening. The success was more or less zero. Their advance guard made it into Lowicz but left it again just as fast. If our divisions could have stayed together, we would have been able to move forward much faster. As it was, the Russians were able to bring together all their available forces in Lowicz, and they pulled out only after they had built fortified positions a few kilometers beyond Lowicz near Skiernewice.

And this is also how they would continue to operate. They would let us get from one of their positions to the next and they would always fortify their positions heavily, as near Lowicz and later near Skiernewice, in order to make it more difficult for us to move toward Warsaw. The heavily forested area here prevented us from advancing more rapidly. It would have taken immense sacrifices to storm in this kind of forest, and our supreme command would never be tempted to make such a sacrifice.

On the 26th via Soppel, Sarry, and Sobota we went back to Zduny and into firing position, but we stayed only until evening. Along the same route we went to Glowno, from which we had originally come only that morning, then to Bielawi, and there, in this nice little town we spent the night. What a sight Bielawi was 2 days later! We rested until 11 a.m. We had arrived only at 4 that morning. Suddenly [there came an] alarm, and [we were out of Bielawi] in a great rush, [though] nothing compared to the alarm near Niedzwietzken.. We had earlier left behind part of our corps in Zduny, consisting of an infantry regiment and one cavalry division. We went into position in Bielawskiewies, not far beyond Bielawi.

At that time our battalion was led by the captain of 5th Battery, Captain Hamann, who had never shown much love for his men and whose abilities in artillery we disputed. How incompetent he really was showed itself in his choice of a position near Bielawi. He ordered 5th Battery, which already had a relatively good position, to move forward even more. When he heard the objection of the battery commander he said, "I will get you into such a mess you'll know neither here nor there."

The first to fall that day was Captain Hamann. But his prediction that he would lead us into a mess came true, because the most difficult day and the one with the most casualties that I was part of during this campaign was that day at Bielawi; and many of the casualties were due to him.

At 6 p.m. the Russians, having 30 times the forces we had, attacked. Miraculously we were able to beat them back. At that point the regimental commander and leader of our detachment, Lieutenant Colonel Müller, rode to our division commander to request help. He could get a promise of help only for the next morning, and so things had to happen the way they did.

At midnight the Russians attacked again. Two guns of ours were in the foremost trench. I have on another occasion described this. Our infantry went back; "went" doesn't describe it though, and "running" doesn't describe it either. Tumbling back is what they did. Our two guns they left behind.

How they got through and back to our battery I do not know. It must have been horrible as one can only imagine. Between the Russians, whose "Stoy, stoy" ["freeze, surrender"] rang in their ears, those brave fellows went. The only one who really knew the way was Sergeant Sengpiel, who, as a matter of fact, was the one who ended up saving our guns. We needed to retreat very, very hastily. When we had retreated 6 to 7 kilometers, we stopped and went into position again. We lost 2 ammunition carriers, 10 men, and 15 horses that difficult day!

Only 3 days later, on 30 November, we took Bielawi back again and even retrieved both our ammunition carriers. But the condition our infantry was in! They had completely lost their nerve. And when on the 30th I was in the foremost trench as an observing officer, I only heard them whine, "It's going to happen to us today just as it did three days ago." Only 14 days later the entire [infantry] regiment, even though surely troops were needed badly, was given 2 weeks of vacation in Thorn.

Lieutenant Wenz, who had led the platoon near Bielawi, deservedly or undeservedly (I leave that in the air) received the Iron Cross 1st Class. Sergeant Sengpiel surely deserved it. At least he received the Iron Cross 2nd Class later on. Generally speaking, it always takes relatively long until the men receive the Iron Cross for which they have been nominated. When you consider that I had been nominated on 7 November and only received it on 10 December, that is a really long time.

* * * *

Early in November Riess and his comrades had been sent south to join the German Ninth Army for their attack into Poland. While their war from August through October had been intense and exhausting, in November it was worse.

Fighting on the Eastern Front was as dangerous a struggle as ever for both sides in November 1914, but especially for the Germans. One strategic or tactical error could lead to disaster for all of Germany. On the Western Front the French and British had stopped the Germans' latest advances. To the southeast their allied Austrian-Hungarian forces had been pushed out of Serbia by local troops and several Russian armies. In Poland the Russians were massing over a million troops who had arrived from their eastern regions: regular infantry and cavalry divisions, more divisions of mounted Cossacks, large numbers of heavy artillery, and divisions of the highly respected Siberian infantry.

The stakes were high for all countries in the war. Each, using the accepted modern doctrine of *total war*, had mobilized its entire population and economy for the war effort. Almost everyone, including even children in each country, was sacrificing something, while millions of soldiers and civilians at the front were losing everything.

Part of mobilizing the public for total war was the need for propaganda to motivate people. Therefore each country's population was being convinced by their government that they were on the side of good and God, they were winning, and they needed to put more effort into defeating the evil enemy. More and more men were mobilized and armed, factory workers poured out thousands of tons of ammunition each day, and everything was rushed to the front.

Each country raced to win the war as the other side's frontline armies grew in strength.

Whereas the Russians had started the war with two armies facing the German's single German Eighth Army, by the end of October the Russians had six armies facing two German armies. The latter were the Eighth and the newly formed Ninth Army, which was commanded by General Hindenburg. Shortly after the Ninth Army was formed, Hindenburg was promoted to Supreme Commander of the East, and the able General August von Mackensen, known as "The Last Hussar," replaced him as the Ninth Army commander.

In the first days of November the Russians were increasing their forces along a roughly north-south line from the Baltic Sea to Hungary, and were massing their rebuilt Second and the Fifth Armies and extra supplies around Lodz, a Polish city of 500,000 people southwest of Warsaw. The Russians were planning a major attack southwest from Lodz through Silesia. From Silesia they thought they would be able to outflank the Germans and attack northwest into Germany's heartland.

German intelligence gathered enough evidence, especially from intercepted radio messages, to allow them to determine the Russians' plans. Outnumbered approximately three to one, the Germans decided to make a surprise attack on the Russian armies massing near Lodz before they were organized into a huge strike force. If they could again destroy the Russian Second Army and then withdraw, they might defuse the Russians' plans. This would be a daring raid on a massive scale.

Mackensen's Ninth Army was chosen to make the attack against the Russians at Lodz while the rest of the German eastern troops engaged the enemy all along the Eastern Front to tie up the Russian units facing them. Using the railroads as efficiently as ever, the entire Ninth Army, with all its horses, equipment, and supplies, was quickly shifted north to the area around and south of the major

German fortress at Thorn. The 1st and 25th Reserve Corps of the Eighth Army transited south to join them.

Sergeant Riess, in the 1st Reserve Corps, which had been part of the Eighth Army since early August, had ridden to Insterburg and then boarded a troop train to Thorn, where his corps became part of the Ninth Army. None of the soldiers had been told where they were going in such a rush, but as they shifted into position near the Polish border east of Thorn, they knew. There they had a few days to resupply, refit, and finally be issued personal firearms—old, used cavalry revolvers. Riess would be in the left (north) wing of the German Ninth Army for the 26-day carnage that today is called the Battle of Lodz.

The fighting began on 11 November with a German attack eastward all along a 110-kilometer front, with 6th Battery on the left flank advancing along the southern bank of the Vistula River. The Russian High Command, listening to the German radio traffic, quickly understood the Germans' plan to hit their Second Army's front and then encircle it from the north. While the Russian Second Army fought and slowly gave up territory, the Russian High Command directed their First Army to march south and their Fifth Army to force-march north to help the threatened Second Army.

Therefore, on 15 November, while the German 25th Corps marched south toward Lodz, the German 1st Reserve Corps encountered the whole Russian First Army. It comprised three corps of mostly Siberian infantry plus Cossack divisions, including a division of the famous Guard Cossacks. Riess was not exaggerating when he wrote that the 1st Reserve Corps in which he served were severely outnumbered by the Russians' best troops. At noon the Russians attacked with artillery, waves of Siberian infantry, and flanking machine gun fire. The attack continued for hours, and though outnumbered three-to-one, the Germans were able to beat the Siberians back and even take about 6,000 prisoners. This was when Riess recorded only, "The 15th of November was a very successful day for us. Fighting was heavy, and the enemy must have lost many men, as did we!"

For the next few days of almost constant fighting, by attacking and withdrawing, maneuvering and firing, and advancing again, the reservists were slowly able to push the Russian First Army back while the rest of the German Ninth Army, just to the south, tried to destroy the Russian Second Army. With the help of an additional division of cavalry, the German 1st Reserve Corps was able to pierce the Russian line, creating a wedge between the enemy's First and Second Armies. On his twenty-seventh birthday, Saturday, 21 November, Riess rode with his battery in pursuit of the enemy. They swung their guns into position and fired on the Russian infantry dug in at Swieryz. That evening the troops were read the following letter from General von Morgen, commander of 1st Reserve Corps:

> Soldiers of the First Reserve Corps!
>
> You have in recent days defeated four Russian army corps by your unparalleled bravery and your untiring perseverance, have captured 20,000 prisoners, taken 31 guns and 40 machine guns. The losses of the fleeing enemy are so great that they will no longer appear in the field.
>
> You have brought about the victorious reversal of the present situation in the Russian campaign.
>
> Your actions have made the name of the First Reserve Corps forever engraved in German history.

It is and will remain the "first" corps.

Emperor and Fatherland are proud of you. On their behalf, I thank all the officers and men for their death-defying duty.

And now continue forward to the complete destruction of the enemy!

With God for Emperor and Empire!

The commanding General
(signed.) von Morgen[7]

The next morning, another massive attack by the Russian First Army continued for hours but was repulsed. Almost immediately after the Russians withdrew from the field, 6th Battery advanced in close support of the infantry as they drove the disheartened Siberians and Cossacks before them. During that week, as the line was slowly pushed east, tens of thousands died and more were wounded on both sides. The field medical teams couldn't keep up with the numbers. Many of the wounded and dead had to be left, frozen on cold ground, as the fighting continued across the fields, through the forests, and into the villages.

At the same time, both sides struggled to bring as many men, artillery, and supplies as possible to the huge conflict. The Battle of Lodz pitted approximately 250,000 Germans fighting against 600,000 Russians for more than three weeks in winter conditions that came early that year. Every hour they fired tons of ammunition

---

[7] *Das Reserve-Feldartillerie*, p. 69.

at each other and used tons of other supplies. Each night, when enemy artillery observers were less likely to see them, columns of horse-drawn caissons and wagons without lights brought ammunition, food, animal fodder, water, and replacement horses and parts to their forward units if they could find them. Often frontline NCOs were sent back at dusk to guide their supply column to the right location. The supply wagons returned to the train depots with wounded men, broken equipment, and empty artillery cartridges.

While the German 1st Reserve Corps fought the entire Russian First Army just south of the Vistula River to keep them away from Lodz, the rest of the Ninth Army made their major strike to surround the Russian Second Army, positioned around Lodz. As part of the strategy, just south of Riess's position, the German 25th Reserve Corps left the battle line and fought its way southeast to the east side of Lodz, almost closing the back door on the Russian Second Army. However, the next day the Russian Fifth Army of 200,000 fresh troops, exhausted after a two-day forced march, appeared from the east and south, encircling the 60,000 men of 25th Reserve Corps. The city of Lodz was then surrounded by intertwined layers of hundreds of thousands of Russian and German troops. At one time several large units of each side were surrounded and fought in all directions. As Churchill wrote in *The Unknown War*, "On this day if a spectator had received a safe conduct to travel from north to south across the armies, he would in a journey of twenty-five miles, have passed through eight separate fighting lines, back to back or face to face."[8]

On 23 November, completely surrounded by the Russians and defending themselves on all sides, 25th Reserve Corps struck north in a fighting retreat. Two regiments of 1st Reserve Corps,

---

[8] Churchill, *The Unknown War*, 260.

including Riess's battery, were ordered south to help them. Riess wrote that 25th Corps saved itself, which is true, but we now know that the southern movement of the two regiments of 1st Reserve Corps forced the Russians between them to withdraw to the northeast, opening the path for the fighting withdrawal of the 25th. Only the Siberian 6th Division remained in the path of the 25th Corps, and the 25th destroyed it before they were again safely in the German line between 1st Reserve Corps and the rest of the German Ninth Army.

Without hesitation, the battle line re-formed to continue the bloody fighting around Lodz. Hundreds of thousands of men were fighting everywhere north, west, and south of Lodz. In the north, just after 6 p.m. on 27 November, swarms of Siberian infantry again stormed the forward positions of 1st Reserve Corps. Riess's two field gun crews found themselves firing canister shot directly into the Russians as the rifle and machine gun bullets bounced off their field guns, flew around them, and hit some of the men and horses.

Riess's two guns were just a few meters behind the front line and fought on in the darkness, only 30 meters from the Russian bayonets as the forward-most German infantry and the other platoon from his battery fell back. This was the day when Riess recorded that the other two gun crews of his battery struggled to retreat to his position, then they all retreated, while fighting at close quarters. Then once again, after a protracted battle, the Russians were beaten back with heavy losses on both sides. Riess gave most of the credit for the other platoon's survival to its Sergeant Sengpiel, while the official regimental history gave the credit to Lieutenant Wenz.

During the fight, Russian shrapnel lodged in the spinal cord of Captain Hamann, the battalion commander whom Riess did not like. The wound was mortal and the next day Captain Krüger, Riess's battery commander, who had been his commander before the war and had proved himself well in the four months of fighting, was given command of the entire battalion.

On the morning of 28 November, while the Russians tried to reorganize after their failed attack, the Germans attacked them to gain a little high ground, then both sides dug in and resumed heavy and light artillery fire while their infantry and cavalry units regrouped. 6th Battery fought from a position north of the road at Drogusza. The next day the battery shifted to the northeast corner of the Drogusza forest and engaged enemy artillery and infantry who were entrenched at Walewice. On 30 November the Russians attacked en masse against their position, again forcing the whole battalion westward for a few hours.

As November closed, the men of Germany's 1st Reserve Corps continued to attack, withdraw, defend, and shift to hold off the much larger Russian First Army on the Germans' north flank. At the same time, fierce fighting continued all along the 100-kilometer front that was centered on the city of Lodz.

Map 7. 1–31 December 1914. Pushing toward Warsaw.

CHAPTER 7

Praying for a Silent Night

1 December 1914–1 January 1915

On 1 December listening patrols were sent to Bielawi at dawn. We waited every second for shots to ring out to bring us proof that the enemy still occupied Bielawi and the surrounding villages. Through observation glasses I was able to see the path the patrols took, but shots could not be heard—the enemy had left Bielawi.

At 10 a.m., after the listening patrols had gone through Bielawi, returned, and reported that the enemy had cleared out completely, we went in and saw what had become of Bielawi in those 3 days. One house was still standing and one chimney, that was all. An enormous heap of rubble was left. And what it had cost us!

We found the charred remains of one of our men. What had happened to the others we did not find out, whether they were dead or prisoners. Of one of our brave men, one I was very fond of, I heard from later in Berlin. He ended up a Russian prisoner. I would like to remember here my faithful combat and sleeping partner Seebe, who during that afternoon, when the platoon had already had losses near Bielawi, volunteered to go forward and help. I hope that the recognition that he truly deserves and which was mentioned at times will not be denied him.

Soon after we passed Bielawi we went into position again and fired at the enemy, who was withdrawing toward Chroslin.

Our next task was to take Lowicz. As mentioned already, it was supposed to have been taken by one brigade before and now 4 corps had to work at it for the better part of

December. On 1 December we withdrew to Milanowo for one night to get some rest. It was the first time in 20 days that permission was given to unharness our horses. On 3 December we went forward from Bielawi to Sobota. Where the journey was leading us, we didn't know.

When we came to Urzyce we were fired at by infantry and artillery along the village road and had to find cover behind buildings. The gunfire did not stop, even after one of our batteries went into position. Consequently, our new battalion commander, Captain Krüger, whom I greatly admire, decided that the entire division should go into position near Boguria-Gorna. In selecting an appropriate position, not an easy task and one that takes some luck, the excellent artillery skills of our captain showed off to the fullest. For almost three weeks, day and night, the Russians shot at us, and not even one projectile came within close proximity of the battalion. We were even given two more heavy batteries plus one howitzer battery, all of which were to shoot at the Russians positioned close to Lowicz.

My time in Boguria-Gorna was rather monotonous. A lot of shooting went on, and some villages that we had passed through before and we passed through again later, when we marched from Boguria-Gorna to Lowicz, had meanwhile been destroyed by gunfire from our battery. Every third night I had to go forward to Urzyce with a platoon to support our infantry if there was to be an enemy attack. But during my entire time an attack never occurred. On 4 December we tried unsuccessfully to shoot down an enemy plane that consequently gave away our artillery position. As a result we were shot at with heavy artillery. Nevertheless the shots never did come close; sometimes they would hit far ahead of us, then way behind but, thank goodness, never at the right spot.

We were immunized here in Boguria-Gorna against cholera. I wasn't there later when typhoid shots were given.

In the afternoon of 10 December, after we returned from the firing position, I was called to the captain. He told me to report at 6 p.m. to the quarters of the major, who had meanwhile returned to our division. I was not aware of any wrongdoing (and if I had been up to something, the captain most certainly would have told me about it) and assumed that this would be about the reward for Suwalki. And I was right about this. When I arrived, the major [Major Kujath], after a lengthy speech, gave me and three others the Iron Cross. When he congratulated us afterward he said to me, "What is Mrs. Riess going to say about that?" As I mentioned at the beginning, this major had been my battery commander during military service, and he had met my wife and me during that year when we were vacationing in Kissingen. The way he handed over the Iron Cross to me was very charming.

On the 13th the Russians shot at and set fire to a farm at Boguria-Gorna and we thought they had located our sleeping quarters. When no other shots were fired that evening, we moved into our old quarters again. One really gets thick-skinned in the face of danger, considering that the Russians knew of our sleeping quarters and yet we lay down there to sleep anyway.

An Austrian motor battery [heavy artillery pulled by trucks] had come to our left flank and sent "greetings" to the occupants of Lowicz. On the 14th the Russians attacked our right flank with such great force that we had to get ready to march. Things stayed quiet in our center position, however.

That evening we saw incredibly heavy fighting on the right flank. During the assault the Russians overran the village and churchyard of Chroslin, and on the 15th our right flank proceeded with a counter-attack with the support of our artillery. We won back the village and churchyard of Chroslin.

This attack on the 14th and 15th, however, had a very specific purpose, totally incomprehensible to us Germans. That 14th as well as on the 15th the Russians suffered incredible losses. Regiment after regiment was chased in front of our machine guns. To what purpose?—The Russians wanted to retreat from Lowicz without being disturbed, and take away their war materiel. People's lives were not important in this. Of course they got what they wanted. Who could assume that the Russians, while launching one attack after another, were actually retreating? Because of these attacks, our left flank and the troops lying directly opposite Lowicz did not even attempt to advance.

On the 16th I wrote in my diary, "All is quiet, as if the Russians were retreating." Listening patrols that were sent out returned with different reports. Some said the Russians were preparing everything for a central attack. Therefore our infantry got ready to be able to meet the attack. The second listening patrol reported, "All is quiet." The third reported, "Constant rattling of wagons."

No shots were fired, something that hadn't happened in three weeks. And when all stayed quiet on the 17th too, new patrols were sent forward during the day and did not get shot at. So we moved forward, away from Boguria-Gorna and the house of the Polish Princess Matka, the name we had given the owner of that awful smelly shack we had lived in for the past 17 days. That same evening we arrived at Lowicz. The troops who had arrived before us had been in combat with the rear guard of the enemy. When we moved in, all had quieted down already. In Lowicz we had good quarters and were able to have a peaceful night. We could see the incredible effect of the Austrian motor batteries, which had torn holes into the ground at the market place and on the streets of Lowicz, 8 meters wide and 3 meters deep.

On the 18th near Sirakowice we went into position. Again, we were dealing with enemy troops that were guarding the retreat of those moving into a very strong position beyond Skiernewice. On the 19th we passed through Skiernewice, where a beautiful hunting lodge of the tsar is located. In contrast to most other Russian towns, this one was relatively clean. Just beyond Skiernewice we got into battle and stayed out in the freezing cold all night. The following day we stayed in the same position and took quarters in Schwabeln. This is a small colony very close to Skiernewice that was founded by German immigrants. To this day the descendants of these Germans speak some German so that at least we were able to communicate with some of them.

It was simply impossible to penetrate the position that the Russians had taken beyond Skiernewice. Trench after trench, wire-entanglement after wire-entanglement. How much we had accomplished in this location up to now I did not know. On the 21st we moved toward the forest another 2 kilometers, receiving strong artillery fire while approaching, and if the Russians had had more guns in that area, things would have been bad for us. We stayed in our good quarters in Schwabeln. Outside, during the day we had to remain in our previous position, as it was impossible for us to go forward and to move across the river and into the forest. Heavy guns of ours were also used at this location. It was interesting for me to watch how these 21 cm guns, because that is what they were, functioned. When they are put into action, they cause an incredible detonation. Despite their large size they are very easy to handle. I am always amazed what technology brings forth.

Our heavy projectiles landed in the enemy trenches and consequently, one enemy battalion wanted to surrender and approached us with raised hands. At that point, machine guns from their own Russian positions aimed at these men and they were mowed down. If this is the spirit by which Russians operate, they won't endure long. But then, what is the value of humans to the Russians?! If men are wiped out today, they are replaced by new troops. Human beings don't figure large.

I came down with a bad cold and was unable to participate in the Christmas celebration out in the forest near our position. Instead I stayed in the house in Schwabeln until 29 December. On the 29th, at 5 p.m. we moved to the left flank of our position where heavy fighting had taken place all day and we took quarters in Bonowino [Borowiny] at 1:30 in the morning. That day I felt very, very ill. When I was about to get going the next morning, at the request of one of my fellow soldiers, a doctor was called, who strongly urged me to call in sick, so as not to ruin my health. Of course I wasn't yet ready to follow his advice, even though I probably had a high fever.

On the 30th I was able to endure. We were supposed to go into position near Bolimow. Instead we were placed beyond a small village as a reserve unit. This was where our commanding general, General von Morgen, made an appearance, greeted our 36th Reserve Division, had every man who had been decorated with the Iron Cross introduced to him, and encouraged us all to continue to do our duty until the enemy was completely crushed. Afterward, we rode to our quarters in Pyarski.

The next morning, when I awoke, I was unable to make any sounds whatsoever and had no choice but to go to the captain. He offered me a leave, so that, as he put it, I wasn't going to be lost to the battery. I was the very first in the division to be granted a leave, and the major without delay approved the request of our captain. And so I rode to Lowicz, accompanied by my faithful Sell, on my dear old sorrel, which I had ridden for the most part throughout this whole campaign. When I arrived at Lowicz after a fast-paced three-hour ride, a transport [train] of the wounded was to leave one hour later that would bring me back to Germany. Through New Year's Eve I was en route to Thorn. I was too worn out to feel excited about returning home. But when I came across the German border again near Wloclawek, I had a warm feeling: you are returning home!

For many days to come, after I had returned home, I heard shots constantly, and if something fell to the floor or somebody spoke in a loud voice, I got startled, a reaction I did not have out in the field, but my nerves were on edge.

At 11 a.m. on the 1st I arrived at Thorn. On the way I saw huge transports moving out to the front. And everywhere this sureness of victory. One had to conclude: men marching out with such enthusiasm would surely succeed.

I left Thorn by midday and arrived in Berlin in the evening. What can I say about my arrival in Berlin? However much I could say about the supreme happiness I felt when I returned home, it would still not be enough to describe it fully. I was simply truly home—after five months of heavy combat, tiring marches, and many, many deprivations. How many of those comrades I went out with, I spent days and nights with, are still left? Few, very few, because these battles we fought in together were followed by many, more difficult ones. And of those who didn't die, get injured, or wind up being taken prisoner, many left due to illness, as I did.

When my leave was almost over and I was about to return to my battery, an old ailment of mine, one I had already suffered from before the war, made an acute reappearance. It was therefore impossible for me to participate further in this great struggle, at least for the time being.

To summarize what I have written about my war experiences, I can say that the troops are in excellent spirits and will easily overcome all hardship and exhaustion while advancing. If for one reason or another it is necessary for the troops to retreat, they will have but one goal in mind: Advance again!

A war fought with the troops we have and under such leadership cannot but result in the great victory we all envision. Seeing at railroad stations today the kind of enthusiasm with which our young troops are heading out, one can only say with great confidence:

Dear fatherland, maintain your calm![9]

****

As sick as he was, Werner Riess was lucky to be alive and at home. He had lived through five months of warfare. Many of his comrades

[9] Translator's note: At the beginning of the First World War the patriotic song "Die Wacht am Rhein" was very popular in Germany. It was not the national anthem. The line quoted here is from that song.

were already dead or badly wounded. The early winter had killed many soldiers, yet Riess had survived the nights of cold and wet bivouacs in the open. In addition to the warfare and weather, cholera already was spreading quickly in war-torn Europe, sickening and killing thousands.

In early December, the German Ninth Army had caught the Russian Second Army by surprise near Lodz and survived the counter-attacks of the Russian First and Fifth Armies. The Germans' original plan was then to withdraw. Instead, the Germans continued to fight their way northeast, toward Lodz and then Warsaw. They had achieved forward momentum in this theater of the war and General von Mackensen chose to continue to fight eastward while the Russians were still off balance. After four months of heavy fighting, the general's troops and officers had learned much about modern warfare. They were hardened, efficient, and well-armed. Many of the Russian units they faced were well trained and equipped, but most were new to the front line. Mackensen estimated that soon the Russians at the front would also be veteran fighters. Though it may not have been part of his original plan, this was the time to keep up the assault toward Warsaw.

In the December advance, the1st Reserve Corps continued to be on the left (north) wing of the German Ninth Army, bearing the brunt of the fighting against the entire Russian First Army. Riess, still in command of two gun crews in 6th Battery, was fighting almost every day: everything from artillery duels at a distance to close fighting against Siberian infantry.

As they fought eastward the Germans quickly rebuilt and enhanced the railroad system behind their front lines. Where no regular railroad existed, they built temporary narrow-gauge lines for very small locomotives and cars. Often within two days of German forces taking new territory, trains could arrive nearby with supplies and new troops. They then transported the wounded and ill soldiers west to German hospitals and took empty containers and brass artillery cartridges to the factories. Their transportation system was extremely important to the Germans on the Eastern Front, as they were severely outnumbered.

The German officers knew that men and horses needed good food, ammunition, medical supplies, and replacements to function efficiently and move quickly from one hot spot to another. As one example pertinent to Riess's memoir, the German reserve field artillery units were seriously outnumbered by the Russian field artillery. Therefore each gun crew needed to fire much ammunition per day to try to reach parity with the enemy. Fortunately for them, the enemy's supply system was not sufficient, so that Russian artillery crews needed to make careful use of their limited ammunition.

Early in December the 1st Reserve Corps was given the tasks of battling their way through Bielawi and then east along the southern bank of the Bzura River. Here as elsewhere, Riess's commanders ordered the field artillery to shift often, sometimes a few times a day, to support an attack or defend against one. December started with 6th Battery firing on the withdrawing Russians' rear guard, which was entrenched in a north-south line roughly centered on Bielawi. The next day the enemy retreated east with the Germans at their heels, firing when they could see a target.

In charge of two of the four guns of 6th Battery, Riess fought through Bielawi, then north to cross the Bzura River at Sobota, then east, and finally north to a hill southeast of Boguria-Gorna. Here his commander chose a good position from which they could fire on the Russian positions to their east and south. They were ordered to remain there, so they dug in and for the next few days fired in support of the infantry when the Russians attacked. Every third night 6th Battery took its turn in the front trenches a bit to the south, to support the infantry line at Urzyce with point blank firing if the Russians launched a night attack at them.

From Boguria-Gorna Werner sent Gertrud the following verses, which rhyme in German and are alternative lyrics to the then century-old Austrian Christmas carol, *Stille Nacht* (*Silent Night)*. His original German version can be found on page 260.

A Little Christmas Poem

Oh, silent night, oh, holy night
Hearts beat faster,
Pains of yearning burn hotter
Of one, of only one I think!

So far away from here is she
Within me only lives the thought of her!
Lord, when will you give us peace?
When, again, will she be with me?

Oh, silent night, oh, holy night,
You, who bring peace on Earth,
Will there be peace, golden peace?
Give it, God, in your power!

South of 6th Battery the fighting for Lodz continued until 6 December, when the Russians, temporarily low on ammunition, withdrew to a new heavily fortified line northeast of the city. While they secretly prepared to launch another major counter-attack, the Russians left strong rear guard units to delay the Germans' advance into the city. Already more than 100,000 men had been killed or gravely wounded in the battle for Lodz.

After the Germans moved into Lodz, the fighting continued for weeks in the farmlands and forests between Lodz and Warsaw, with the front line shifting back and forth. General Mackensen modified his plan of a massive, quick raid to keep the Russians out of Silesia, switching to a longer-term strategy to push the enemy northeast, with their backs to Warsaw. Holding General Mackensen's left flank, General von Morgen's 1st Reserve Corps had succeeded better than expected. Therefore Mackensen ordered 1st Reserve Corps to attack farther east and to capture the city of Lowicz whenever they could.

Following two weeks of fighting off Russian attacks at Boguria-Gorna (northeast of Bielawi), 6th Battery supported the infantry pushing east along the southern bank of the Bzura River toward Lowicz. The Russians made a fighting withdrawal from Lowicz, and the German troops fought their way in, then paused only for the night.

From Lowicz they were ordered to advance southeast to help capture the small city of Skiernewice. As the Russian czar's palace in Skiernewice was an important political target, the German generals expected the enemy to defend it fiercely. However, again the Russians sacrificed only a small rear guard force to slow the Germans. Meanwhile the Russians prepared serious earthworks and barbed-wire defenses on both sides of the Rawka River.

The Rawka, a tributary of the Bzura River, was only about 10 meters wide. It was approximately two kilometers northeast of Skiernewice. Neither Riess nor the regimental history mentioned ice, but in that very cold December of 1914 the river probably was at least partially frozen.

The Rawka River proved to be a formidable defensive position for the Russians, and they fortified its eastern shore with infantry, machine gun, and artillery emplacements. The men of 6th Battery first established a good position on the edge of the forest just outside Skiernewice, entrenched, and fired at enemy positions near the river-crossing village of Ruda. The author of the German army's official regimental history, usually only dryly presenting a chronology of the facts, described in detail the forest's exceptional beauty: the wonderful tall trees, green moss, and more. Today it is part of the Bolimowski Park preserve.

The men on both sides of the river, and really everyone in Europe, were shaken as Christmas approached. It was the most important family holiday in Europe. They had thought the war would be intense but brief. Almost everyone in every country had repeated the mantra, "The troops will be home by Christmas." By mid-December it was clear that this was far from the truth, as each day more and more men left home to fight on the Eastern or Western Front. In late December millions of individuals were at the front lines, far from their families; and hundreds of thousands of them were already dead or badly wounded. The soldiers, their wives, children, parents, sisters, brothers, and other loved ones were sad and anxious as Christmas approached.

This was the week when Riess became so ill from what he described as a cold with a strong fever that kept him bedridden. Considering his exhaustion, fever, and eventual diagnosis, Riess's symptoms suggest an infection or inflammation due to his gallbladder problem. While he lay ill in a small house near Skiernewice, most of the German Ninth Army enjoyed a short respite from fighting on 24 and 25 December. They posted extra sentries to forewarn of any Russian attack, and then held many small Christmas Eve parties near their frontline positions—in the forest, in the towns and hamlets, and along the roads. They erected and decorated Christmas trees with whatever they had at hand, sang Christmas carols, raffled off presents from the Red Cross, and opened the few presents from home that had made it to the front. Some gifts were from German school children, who pooled their pennies to buy presents for alumni of their schools who were serving as soldiers.

The Russians, dug in on the opposite side of the narrow river, were also celebrating Christmas as best they could, but then they began an artillery barrage that night. Major Kujath, Riess's commander, was struck and knocked down by an exploding shell casing. It had evidently ricocheted off a rock, as it only bruised him badly. Meanwhile General von Morgen's staff was planning to advance in a few days, and the regiment prepared to attack across the Rawka River. They shifted 6th Battery north a few kilometers to Pyarski for a day of resupply and repairs. That was the last march Riess made with the men of his battery as he became so ill that his commander sent him home on sick leave.

Exhausted from his illness, Riess arrived home in Schöneberg, Berlin, on the evening of 1 January, to find his wife also very ill with a high fever from acute appendicitis. She could hardly move and he collapsed into her bed.

Early the next morning in Poland, his comrades of 6th Battery closely engaged enemy positions as the entire 1st Reserve Corps advanced across the Rawka River and, after an extremely bloody battle, pushed the Russians back from their trenches.

* * * *

Gertrud Riess wrote a second letter to her unborn son two weeks after her letter to him quoted in chapter 1. Again, this is a translated excerpt:

> November 1915
>
> Shortly before Christmas 1914, I had another bad attack of an old ailment—appendicitis—that had troubled me many times before, also in these months on my own. Christmas presents for my dear ones away had been finished way ahead in November already, 3 of which were for my dear husband. One of these parcels contained a small Christmas tree that my husband never received.
>
> A letter from him to me to be opened on Christmas Eve had arrived a week ahead. It contained a dear little Christmas poem, a few lines written in a mood of yearning, yearning for home and yearning for peace. [She wrote and sent him the following lines]:
>
> Silent night, holy night,
> All is calm all is quiet
> Only the holy couple keeps watch
> Over the happy, curly haired little boy
> Sleep in heavenly peace, sleep in
> heavenly peace!

Little children cheerfully singing,
Their sweet sounds are heard
Here under brightly lit candles
Bringing joy to the hearts of all
In the holy night, in the holy night

Silent night, holy night,
Yearningly the beloved stays awake
Thinks wistfully of home, of home,
Standing alone now in battle
In the holy night, in the holy night.

Silent night, holy night,
But see! Come see!
Future dreams are slowly unfolding,
I see us in peacetime—happily
United in love, united in love!

What I had seen as a future dream, when I wrote down these lines that day, became truth, glorious truth. My dear husband returned to me on January 1st, 1915. This is what happened:

On Christmas Eve I first felt a bit better, and I got up. In the evening the pain felt stronger and the fever returned. On the first Day of Christmas a nurse had to come; I was delirious at that point and couldn't figure out who that person was who was constantly in my presence. Apparently when she first appeared I screamed every time she came close to my bed.

Apart from my doctor a professor was also consulted and both, together with my father-in-law, recommended that I be operated on immediately. I refused steadfastly to be operated on during my husband's absence and without his knowledge. The New Year arrived. I was not allowed to receive visitors and the nurse had told me so many details about the New Year's preparations at her nurses' boarding house that I decided to send her there toward evening, so that she could attend the festivities. She was going to return soon—but she didn't.

As the New Year's bells starting ringing, I lay in bed feverishly and nobody was with me. The bells sounded muffled to me and I felt terribly uneasy. Suddenly I heard—whether because of the fever, auto-suggestion or in reality—a very clear, cheerful bell, the bell of the New Year. I said to myself: don't listen to these muffled, sad bells, there is another one that you must listen to, it rings clear and full of hope, it speaks the truth to you, it speaks of beautiful days to come in the new year. Listen to that one!

Every second I have described here has stayed clearly in my mind and this strange, dream-like and clairvoyant mood will be unforgettable to me. Late in the afternoon of the 1st of January the doorbell rang. I sent the nurse forward to tell me about it. She returned after some time and I said to her, "Nurse, this has taken a long time, what is the matter?"

She said, "Madam, you will be pleased."

I said quickly, "Is there mail from my husband?"

The nurse replied, "Something much better."

"Is it perhaps even a dispatch from him?"

"Even better, Madam!"

I raised myself a little bit and asked slowly, "Is—he—here?" At the same moment, my beloved husband, who had stood at the door and had heard everything, fell onto my bed.

I will not attempt in the slightest to render my feelings and sensations; it would be completely useless. Everything appeared to me as a confusing dream and for a long time afterward, and sometimes to this day, I am afraid I will wake up.

In the evening of January 1st, my bedroom, that nobody had been allowed to enter during my illness, filled with many people. The news "Werner is here for a few days of vacation" had spread through the family like wildfire. And now everyone came to see him. He had gotten thin, had laryngitis and couldn't speak well, only whisper. He needed to get well again as soon as possible.

The laryngitis was soon better and Werner readied himself to return to the front; a new suit, a thick coat, shoes etc. were acquired, many little presents bought for his comrades, and every evening we celebrated heart-rending goodbyes from one another. Then my dear husband, perhaps as a late reaction to what had happened to him and what he had suffered, perhaps caused by a change in feelings or by all the excitement, had a serious gallstone attack.

Even before the start of the war he had suffered from it, and had been taking the cure in Karlsbad just before the war. He hadn't said anything about this condition at his medical exam before leaving for the war and had endured well the difficult months there. Now he lay down sick, was transported to the military hospital [Vereinslazarett], where I followed him shortly afterward as a new occurrence of appendicitis forced an immediate operation on me. It was a success and everything went well.

After several weeks of treatment, my beloved husband came home. He was temporarily pronounced unfit for active and garrison duty. He voluntarily reported to the commandant of Berlin where he works to this day. In May of 1915 it became a certainty that a long hoped for wish would be fulfilled, I was expecting a child.

Epilogue

From his wife Gertrud's diary we know that Werner Riess seemed to recover quickly from his illness and prepared to return to 6th Battery. He had a new uniform and many small presents for his comrades. But then he was hospitalized with a new gallbladder attack, and during his hospital stay, Gertrud underwent surgery for her appendicitis. In late March or early April 1915, while in the hospital recovering from gallbladder surgery, he wrote his memoir and had two copies typed. In mid- to late April Gertrud and Werner conceived their only child, Herbert.

Because of his recurring illness, Riess was reassign to Army headquarters in Berlin. In November 1915 he was still working there, but the extant records do not provide any clues about what he was doing or anything else about him. Most such records were destroyed in the Berlin fires of World War II.

From the pages of his memoir one can see some evidence of what today is called post-traumatic stress disorder (PTSD). Survivor's guilt and constantly being on guard are typical symptoms. As Riess wrote:

> What is so depressing every morning when we get up and every night when we go to bed: we continue to lead our old quiet life—and out there the world is bleeding!
>
> How many of those comrades I went out with, I spent days and nights with, are still left? Few, very few, because these battles we fought in together were followed by many, more difficult ones.
>
> After I returned home, I heard shots constantly, and if something fell to the floor or somebody spoke in a loud voice, it startled me, a reaction I did not have out in the field, but my nerves were on edge.

We do not know if PTSD was a major problem for him and his family, as little is known about him except for his memoir and some pages in Gertrud's diary. In November 1915 she wrote that they lived happily and had a son in January 1916. However, in 1918 they were divorced.

To date (2020) I have found no written record of Werner Riess's death. Some family lore has it that he died in 1918 in the influenza pandemic that killed so many people worldwide. A different story is that he began to drink heavily, gambled, and after the divorce committed suicide. Either is possible. The influenza pandemic killed millions of people—including a high percentage of people in their twenties. The story of his downward spiral to dissolution and suicide fits tens of thousands of First World War soldiers, in all the countries involved, whose lives were destroyed by post-traumatic stress disorder.

After Werner's death, Gertrud Riess married Kurt Danzinger in 1920. The family of three moved to Heidelberg and in 1923 Gertrud gave birth to another son, Peter. In the early 1930s Hitler's followers had begun systematically promulgating anti-Semitism throughout German society. Though Gertrud and her family were practicing Lutherans, they had ancestors who were Jewish. Victimization was closing in, even for those whose family background was only partly Jewish, therefore the family immigrated to the United States of America in the early 1930s.

Guido Riess (Werner's father), Herbert Riess (Werner's son), and Louis Grumach (Werner's father-in-law), c. 1917.

The Following Section Is A Scan of Werner Riess's Original Memoir As typed in April 1915

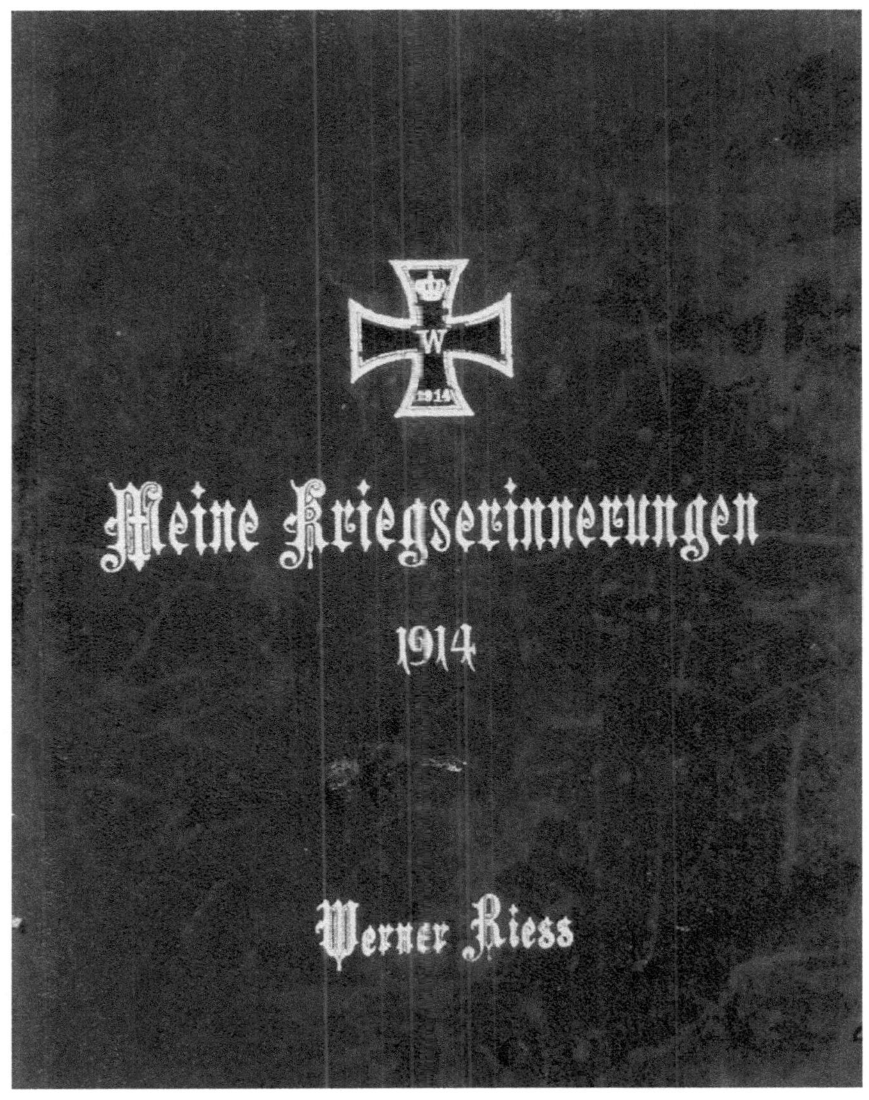
W
1914
Meine Kriegserinnerungen
1914
Werner Riess

Meinem lieben Schwiegervater
Werner Rieß
Wiesbaden 2/5. 1915

M e i n e

K r i e g s - E r i n n e r u n g e n

1 9 1 4 .

W E R N E R   R I E S S .

-1-

Am 28. Juni 1914, es war ein herrlicher Sommertag - in Hamburg wurde das Deutsche Derby gelaufen, bei dem ich ja seit Jahren niemals fehlen durfte - wurde der Grundstein zu dem grossen, grossen Weltkriege gelegt, der über uns alle, über Freund, wie Feind, so unsagbares Unglück bringen sollte und der, während ich heute in der Mitte des April 1915 diese Aufzeichnungen machen will, noch in vollem Gange ist, der an Stärke gewiss noch nicht abgenommen hat, sondern eher noch zunehmen wird, denn Entscheidungen von grosser Tragweite, oder wenigstens von entscheidender für die eine oder andere Partei, sind noch nicht gefallen. Aber unsere Lage ist eine gute und der endgiltige Erfolg wird uns wohl ziemlich sicher sein.

Damals in den August-Tagen, als wir immer von neuen Feinden hörten, immer wieder hörten, dass neue Nationen, von denen wir zum grossen Teil angenommen hatten, dass sie in einem Weltkriege unsere Partei ergreifen würden, sich auf die Seite unserer Gegner gestellt hatten, haben wir uns draussen gefragt: Wie kann Deutschland allein alle die Völker, die sich ihm entgegenstellen, niederringen ? - Aber bis zum heutigen Ta-

-2-

ge haben wir Deutschen gezeigt, dass wir es können werden, dass wir allen Feinden widerstehen können durch unseren Mut, unsere Tapferkeit, das Ertragen von den grössten Strapazen, unsere Vaterlandsliebe und das unbedingte Vertrauen zu unserer obersten Heeresleitung, das jeder von uns hat.

Damals schon, an jenem denkwürdigen 28. Juni, als auf dem Rennplatze plötzlich die Nachricht herumschwirrte, der österreichische Thronfolger und seine Gemahlin seien einem schändlichen Mordanschlage zum Opfer gefallen, war man allgemein der Ansicht, dass dies zu den weittragendsten Verwickelungen führen würde. Daran hat aber wohl keiner gedacht, dass das die Ursache zu dem furchtbarsten Weltenbrande werden könnte, den je die Weltgeschichte zu verzeichnen hatte. Anfang August wusste man es: wir stehen vor dem Beginn des grossen, seit Jahren gefürchteten und doch stets noch vermiedenen Weltkrieges, von dem wir uns alle keine Vorstellung machen konnten. Aber er war nun einmal da und der Deutsche Kaiser, der seinem Lande immer den Frieden hat erhalten wollen, der bei jeder Gelegenheit seine Friedensliebe gezeigt hat und dessen höchster Wunsch es sicher gewesen ist, als der Friedenskaiser einst in der Geschichte der Deutschen gefeiert zu werden, er sollte der Kaiser werden, der Deutschland in den gewaltigsten Krieg führen musste.

Wie waren die Tage, kurz vor und kurz nach der Kriegserklärung, hier in Berlin ! Es war eine Begeisterung, dass man das sichere Gefühl hatte: wir müssen siegen ! Aber auf der anderen Seite hatten wir alle doch das Gefühl: was lassen wir zuhause zurück ! - - Und schön waren die Tage vor meinem Hinausziehen gewiss nicht. Es war mir doch bitter schwer geworden. Viel, viel saurer, als ich es mir jemals selber habe eingestehen wollen. Man wusste doch damals noch von nichts, wusste nicht, wohin man kommt, wusste nicht, würde man Menschen, mitfühlende Menschen draussen finden, die auch ein Interesse an einem zeigen würden. Und würde man alle die wiedersehen, an denen das Herz doch so innig hing ? -

Also am dritten Mobilmachungstage, es war der 4. August 1914, musste ich von Hause fort. Es war schwer - bitter, bitter schwer. Auf meiner Kriegsbeorderung stand, dass ich mich nachmittags um 5 Uhr auf einem bestimmten Schulhofe stellen musste. Und schon da merkte ich, dass das Herrenleben doch wohl für eine ganze Weile ein Ende haben würde. Wie der erste "Kamerad" - er hatte keinen Kragen um und ging in Pantoffeln - mich jovial auf die Schulter klopfend fragte: "Sage mal, wo sollen wir uns denn hier zusammenfinden ?" - Aber ich bin ja immer ein Mensch gewesen, der sich allen Situationen anpassen kann, und so zogen wir beide - ich habe ihn später niemals wiedergesehen, weiss auch nicht, wer er gewesen und wohin er gekommen - auf unsern Schulhof, von wo unser Transport abging.

-4-

4. August: Berlin-Marienwerder ! Von abends um 8 Uhr bis zum nächsten Abend um 9 Uhr; (ich glaube, ein Schnellzug fährt die Strecke in normalen Zeiten in ungefähr 8 Stunden,) aber in jedem kleinen Nest wurde Halt gemacht. Ueberall wurden Erfrischungen für die ins Feld Ziehenden gereicht, und ich möchte beinahe sagen: zuviel des Guten war es, was geboten wurde, denn man musste schon wirklich einen sehr gesunden Magen haben, um unbeschadet seiner Gesundheit den Verlockungen zu folgen, mit denen die teils schöne, teils weniger schöne Damenwelt Westpreussens einen beglücken wollte.

Ich fuhr in einem Wagen 2. Klasse mit dem Transportführer, einem Stabsarzt, einem Apotheker und zwei Vize-Wachtmeistern zusammen. Was aus den Offizieren geworden ist, habe ich nicht erfahren, die beiden Vize-Wachtmeister sind aber zu mir in die Abteilung gekommen. Der eine, Vize-Wachtmeister Schröter, war es noch, als ich fortging, der andere, Vize-Wachtmeister Masch, seines Berufes Opernsänger in Stettin, hat, nachdem er von seinem ihm so "zugetanen" Batterie-Chef, Herrn Hauptmann Hamann, zum 25. Corps versetzt wurde, beide Beine und einen Arm verloren. Ich erwähne das schon hier an dieser Stelle, weil es eben gerade einer von denen ist, mit denen ich hinausgezogen bin und ich hier sagen will, wie sehr man seinem Schöpfer danken muss, dass man jetzt hier mit gesunden Gliedern sitzen kann, während er, der gewisslich nicht in den glänzendsten Verhältnissen lebte, mit seiner jungen, eben kriegsgetrauten Frau, wenn er überhaupt noch am Leben ist, ein Dasein führen wird, das doch so bitter hart ist, so unerträglich, wie es schlimmer garnicht gedacht werden kann. Aber sicher ist er nur einer von Millionen, die dasselbe Schicksal haben, und das ist es, was uns immer wieder von neuem deprimiert, jeden Morgen, wenn wir aufstehen, jeden Abend, wenn wir zu Bett gehen und wenn wir unser altes Leben weiterführen. Wir leben unseren ruhigen Tag weiter, - und draussen blutet die Menschheit !

Ich will nun aber wieder zurück zu meinem Thema. Also am 5. August trafen wir in Marienwerder ein und hatte ich dort, ich möchte sagen, das erste Glück in diesem Feldzuge, das mich ja, Gott sei Dank, nicht verlassen hat. Der Major, der unsere Abteilung führte, war mein alter Hauptmann Kujath, der mir schon in meiner Einjährigenzeit so besonders zugetan war. Er erkannte mich sofort, und es war wirklich angenehm, zu wissen, dass man einen Menschen da hat, der doch schon einiges Interesse für einen gezeigt hatte und weiss, aus welchem Hause man kommt. Am Abend wurden wir noch in Batterien eingeteilt und dann durften wir Reserve-Unteroffiziere in ein Hotel gehen. Als ich das Hotelzimmer betrat, dachte ich bei mir: O Gott, so fängt also der Krieg an ! Das Primitivste vom Primitiven, aber wie das Schlafzimmer in einem Fürstenschloss wäre es mir *vorgekommen, wenn ich es später gehabt hätte. Im ganzen Feldzuge, den ich mitgemacht habe, vom 4. August bis zum 31. Dezember, war es mir sicherlich nicht zehnmal beschieden, in einem so mit "allem Komfort" ausgestatteten Zimmer meine müden Glieder in ein richtiges Bett zu legen, mich am Morgen an einem richtigen Waschtisch zu waschen und an einem Tisch zu frühstücken.*

Am 5. nachmittags, nachdem wir in Marienwerder eingekleidet waren, ging es nach Gross-Krebs. Ich hatte mich mit meinem Wachtmeister Martin schon feste angebiedert, und er nahm mich mit sich in sein Quartier, wo für mich in einer Dachkammer neben Aepfeln, Birnen, Mäusen und Ratten Betten hingelegt waren. Aber es war wirklich wunderschön, denn man konnte sich ja ausziehen ! Das Essen und die Verpflegung waren tadellos und die Zeit in Gross-Krebs war sicherlich nicht die schlechteste, die ich durchgemacht habe.

Am 10. war alles fix und fertig ausgerüstet; wir fuhren mit unseren Geschützen von Gross-Krebs nach Marienwerder und wurden dort nach Nordenburg verladen. Am 11. gegen Abend waren wir in Nordenburg, und nach einem Marsch von ungefähr 2 Stunden kamen wir in unser erstes Biwak nach Wilhelmssorge. Am nächsten Morgen ging es weiter von Wilhelmssorge nach Bidaschken. Bidaschken war ein Nest von ungefähr 30 bis 40 Häusern, wo ein paar tausend Soldaten untergebracht waren. Man kann sich unser Quartier also ungefähr vorstellen. Zu essen

-7-

hatten die Leute nichts mehr, und es fing dort schon der eigentliche Krieg an. Aber zum Glück blieben wir nicht allzu lange da und am 14. waren wir in Paulswalde. Dort habe ich es wirklich gut gehabt - bei der Familie Stadthaus, derer ich freundschaftlichst gedenke; sie haben mir dort wirklich viel Gutes erwiesen und ich weiss, dass die alte Frau Stadthaus immer sagte: "Wir wollen doch alles geben, was wir nur haben, wenn nur nicht die Russen kommen." Und wir in unserem sicheren Siegesgefühl, wir haben ihnen gesagt, es sei doch ausgeschlossen, wir wären ja so stark und es wäre so unmöglich, dass uns die Russen schlagen könnten. Und wenige Tage später, am 20. schon, hatten die Russen Angerburg und auch damit Paulswalde in ihrem Besitz. Und als ich mich später bei Flüchtlingen aus Paulswalde erkundigte, ob sie was von meinen alten, lieben Quartierwirten gehört hätten, sagten sie mir nur, dass auch sie hätten fliehen müssen, wie eben alle aus der Gegend. Hoffentlich haben die russischen Horden ihnen ihr schönes Besitztum nicht völlig zerstört.

In Paulswalde wurde viel Dienst gemacht, Geschützexerzieren, Richtübungen usw. Mir war damals die Gefechtsbagage übergeben worden, sodass ich mit dem eigentlichen Exerzieren wenig zu tun hatte. Um so häufiger war ich in Angerburg, machte dort Besorgungen für die Batterie und habe dadurch noch einige ganz schöne Tage verlebt, und wenn ich wieder in mein Quartier zurückkam, wurde ich recht verzogen. Ich hatte dort

2 grosse Zimmer, während vier Infanterie-Offiziere zusammen in einer Dachkammer wohnen mussten. Gewusst haben sie allerdings von meinen 2 Zimmern nichts, denn sonst weiss ich, dass sie mich höllisch schnell an die Luft gesetzt hätten.

Am 19. frühmorgens gab es Alarm, und wir rückten von Paulswalde fort, um uns in einer Stellung zu verschanzen, die einige Kilometer von Paulswalde gelegen war, bei Sobiechen. Es hiess, wir sollten dort mehrere Tage bleiben, und deswegen fuhr ich gegen Abend wieder nach Angerburg, um noch Einkäufe zu erledigen. Als ich nachts in die Nähe von Sobiechen kam, empfing mich einer unserer Trompeter und erzählte mir, dass die Batterie schleunigst habe abrücken müssen und dass wir versuchen sollen, entweder mit ihr oder mit unserer grossen Bagage zusammenzustossen. Letzteres gelang mir auch. Die nächstfolgenden Tage blieb ich bei der grossen Bagage, da unser Corps in schwerem Kampfe war. Die wildesten Gerüchte tauchten auf, man hörte von einem grossen Siege, eine halbe Stunde später war es eine schwere Niederlage. Ersteres war es aber, wie sich nachher herausstellte, bestimmt nicht, denn am 23. waren wir wieder bis nach Nordenburg zurückgenommen worden. Der Führer unserer Armee, von Prittwitz und Gaffron, hatte sich auf den Standpunkt gestellt, uns bis an die Weichsel zurückzunehmen und das schöne Ostpreussen ganz den Feinden zu überlassen. Aber unser Kaiser hat es anders gewollt. Gott sei Dank ! Ich entsinne mich, dass es am 23. hiess, wir haben einen neuen Führer bekommen, einen Herrn von Benekendorf. Wer sich hinter dem Namen dieses Herrn von Benekendorf verbirgt, das haben wir erst einige Tage später erfahren, als unser grosser Sieg bei Tannenberg gemeldet wurde, denn heute wissen es wohl auch die wenigsten, dass unser grosser Generalfeldmarschall von Hindenburg früher als seinen Hauptnamen den Namen von Benekendorf getragen hat und sein voller Name eben von Benekendorf und von Hindenburg ist.

Ich will nun aber beim Thema bleiben: am 22. abends war unsere Bagage in Nordenburg. Es wurde dort erzählt, von unserer Batterie sei keiner übrig geblieben, von der 5. Batterie nur der Hauptmann und der Wachtmeister usw. Was stimmte wirklich ? Unsere Batterie hat keinen Mann verloren, die 5. Batterie einen, der vom Pferde gefallen war. Es wird nirgends so viel geschwindelt, wie im Kriege. Aber froh waren wir doch, als wir alle unsere Kameraden wieder mit heiler Haut ankommen sahen.

Am 23. sollten wir einen Ruhetag in Nordenburg haben. Aber schon um 11 Uhr gab es Alarm und unweit von Nordenburg bei wirklich schönem Wetter Biwak. Am 24. schrieb ich in mein Tagebuch: "Rückzug, Biwak." - Wie weit wir damals zurückgegangen sind, an dem Tage, kann ich nicht sagen. Ich weiss nur, dass wir am 25. wieder vorgenommen wurden, und zwar sollte unser Corps dem 20. Corps zu Hilfe kommen, das in schwerem Gefechte war.

Schon am Mittag war die Nachricht verbreitet, dass das 20. Corps einen schönen Sieg über den Feind errungen habe. Das war der erste Tag der gewaltigen Schlacht von Tannenberg, an welchem wir aktiv noch nicht eingreifen durften. Um so schwerer sollten wir aber am nächsten Tage ins Feuer kommen.

Am 25. noch war die Nachricht bei uns verbreitet worden, dass Italien an Frankreich und Russland den Krieg erklärt habe. Heute, wo ich erst wieder in den Zeitungen von grossen Demonstrationen der italienischen Studenten gegen die deutschen Professoren gelesen habe, kann ich nur darüber lächeln. Damals haben wir kolossal Hurrah auf Italien gebrüllt. Schade drum ! -

Also am 26. August sollte auch unserem Corps beschieden sein, an der grossen Schlacht bei Tannenberg mit teilnehmen zu können, der grössten Schlacht, die bis zum heutigen Tage dieser Krieg gezeitigt hat. Wir standen an jenem 26. bei der Oberförsterei Sauerbaum. Um 10 Uhr vormittags war weder etwas zu sehen, noch ein Schuss zu hören, nicht einmal, was wir aus den Tagen vorher schon kannten, das ferne Donnern der Geschütze. Alles war ruhig, und keiner von uns hatte eine Ahnung, dass wir in wenigen Stunden in einem so schweren Gefecht sein würden. Plötzlich kam das Kommando: Aufgesessen ! und im Trabe ging es vorwärts. Wenige Kilometer durch den Forst auf Wegen, die keine Wege waren, zwischen Bäumen und Sträuchern hindurch gings im flotten Trabe. Plötzlich neben uns die ersten Granaten. Es war nun das erste Mal, dass ich ins Gefecht kam. Ich weiss heute selbst nicht mehr, welches meine Empfindungen im Augenblick gewesen sein mögen. Die Batterie war schon vor und ich mit meiner Gefechtsbagage einige hundert Meter hinter ihr und versuchte nun, die Protzen zu erreichen. Als ich an eine Lichtung kam, bekamen wir Feuer, sodass ich meine Leute absitzen lassen musste und sie hinter die Wagen stellen liess. Zu sehen war von der Batterie nichts. Ich schickte deswegen unsern Fahnenschmied aus, der mir melden sollte, wohin sich die Protzen in Deckung begeben hätten. Er ist nicht wiedergekommen. - Ich habe später erfahren, dass er schwer verwundet in ein Lazarett gekommen ist. Dann plötzlich erscholl der Ruf: Sanitäter ! Mein braver, biederer, meist betrunkener Sanitäter Cibulski machte sich also auf den Weg, um dem Ruf zu folgen. Aber wenige Meter von den Wagen entfernt erreichte auch ihn schon das Schicksal. Vier Kameraden schleppten auch ihn schwer verwundet fort. Das war nun ein netter Anfang für mich, denn ich hatte niemand weiter bei mir, war der einzige Unteroffizier, hatte auch nicht den Mut, noch einen meiner Leute fortzuschicken, weil ich bei dem enormen Feuer annehmen musste, dass jeder dabei verloren war. Ich tat nun das Klügste, was man wirklich in dieser Lage tun konnte, unbewusst. Ich blieb stehen und wartete, bis sich das Feuer etwas gelegt hatte. Es war ein wahnsinnig heisser Tag. Unsere Führer, noch völlig unerfahren, sind durch die Sandwüste gefahren, und bei jeder Staubwolke wussten die Russen, wo wir standen, und da meine Stellung die einzige Zufahrtstrasse zur Gefechtslinie war, so befand ich mich wirklich in einer Situation, die nicht beneidenswert war. Nachdem für eine halbe Minute das feindliche Feuer ausgesetzt hatte, liess ich aufsitzen und in gestrecktem Galopp an einen andern Platz fahren, der mir geeigneter erschien, und zu meiner Freude sah ich meinen Wachtmeister mit unseren Protzen stehen. Wäre ich aber nur einen Augenblick später dort angekommen, dann wäre wohl nicht viel Lebendes von uns übrig geblieben, denn auch den Staub meiner Wagen hatte die feindliche Bande bemerkt und auf diese Stelle ihr fürchterliches Feuer gerichtet. Als ich zu unseren Protzen kam, wurde gerade ein Verbandplatz der Infanterie aufgeschlagen, und ich habe das erste Mal sehen können, ein wie unendlich grosses Unglück der Krieg ist. Immer wieder neue Verwundete, Freunde und Feinde, Offiziere und Mannschaften, wurden eingeliefert. Es war entsetzlich ! Aber kaum eine Viertelstunde konnte der Verbandplatz an dieser Stelle bleiben, dann musste er abgebrochen werden, weil der Feind gerade diese Stelle mit seinen Geschossen bestrich. Gewusst hat keiner von uns, was eigentlich los war, wie die Lage war, ob gut oder schlecht für uns. Am Abend, bei eingebrochener Dunkelheit, wurde das feindliche Feuer geringer, und als unsere Protzen an die Geschütze herangeschoben wurden, sagte uns unser Hauptmann, der beim Major war, wir hätten allem Anschein nach einen grossen Sieg errungen. Warum und weshalb *war uns aber garnicht klar. Ich entsinne mich an dem Nachmittage an eine furchtbare Situation, die uns mitgeteilt wurde. Eine unserer schweren Artillerie-Batterien musste auffahren, und zwar in einer Entfernung von ca. 5000 Metern, und den Feind beschiessen. Unsere prachtvoll stürmende Infanterie kam mit dem Feind in engste Berührung. Und der Erfolg war, dass die ersten Salven, die unsere schwere Artillerie abgab, auf unsere eigenen Truppen fielen. Wie deprimierend das sein muss, vorn vom Feind und von hinten von eigenen Truppen beschossen zu werden, - obgleich ich es, Gott sei Dank, niemals mitgemacht habe, ich kann es mir aber vorstellen.*

Also am Abend bezogen wir Biwak. Ich hatte noch vorher einen kleinen Verwundeten-Transport, aber nur ganz leicht Verwundete, zum Verbandplatz zu bringen, da ich genau wusste, wo er war, und sah nun diese hunderte und aber hunderte von Leuten auf der Erde liegen, denen noch garnicht geholfen werden konnte, denn es waren auf den einzelnen Verbandplätzen zu wenig Aerzte. Es konnten auch nicht mehr dort sein, unmöglich. Unsere Batterie hatte bis auf meine beiden Leute keinen Verlust. Nur dass unser Offizier-Stellvertreter Glaubke einen leichten Armschuss erhalten hatte, der ihn für einige Tage von der Batterie trennte. Aber noch nicht ganz ausgeheilt kehrte er wieder zu uns zurück. Die Truppen, mit denen wir die Schlacht bei Tannenberg gewonnen haben, das war unsere Elite, das waren die Vorwärtsstürmer, die durch nichts zurückgehalten werden konnten, und man erzählt, dass unsere Infanterie-Offiziere bei dem grossen Sturm am Abend, der die Entscheidung bei Sauerbaum gebracht hat, mit den Revolvern in der Hand unsere Leute zurückgehalten haben, - mit einer derartigen Wut sind sie vorgegangen. Ob es heute, nach 9 Monaten, noch so ist ? ! -

Am nächsten Morgen ganz früh ging es weiter zur Verfolgung des Feindes. Hier sah man 200 gefangene Russen führen, dort wieder 500, mal 2000 und dazwischen immer wieder kleine Trupps von einzelnen. Da erst wurde einem klar: es muss doch etwas Gewaltiges gewesen sein, was in jenen Tagen geleistet wurde.

Den ganzen 27. wurde marschiert, hier und da mal ein paar Schüsse abgegeben, ohne dass wir aber etwas Positives vom Feinde gemerkt hätten. Am 28. waren wir in der Nähe von Allenstein. Ein feindlicher Flieger, der am Morgen die Nachricht nach Allenstein bringen sollte, dass starke Kräfte von uns im Anzuge seien, war am Morgen von unseren Maschinengewehren heruntergeholt worden, und so konnten wir bis auf 2 Kilometer vor Allenstein herankommen, ohne vom Feinde bemerkt zu werden. Das Gros des Feindes hatte Allenstein so wie so schon verlassen, da die Russen sich, wie sich nachher herausstellte, in den Armen unseres 17. Corps wohler zu fühlen schienen, als bei uns. Uebrig geblieben sind ja nicht viele. Ungefähr eine

Brigade war noch in Allenstein zurückgeblieben. Auf einer Anhöhe vor Allenstein gingen wir in Stellung und unsere Infanterie wurde nun auf Allenstein angesetzt. Unsere Artillerie durfte keinen Schuss abgeben, und gar nicht lange dauerte es, da liefen die Russen in hellen Haufen aus Allenstein heraus auf den beiden Chausseen, die aus Allenstein herausführen. Und jetzt fing unsere Artillerie an, zu wirken. Die Chausseen wurden unter Feuer genommen und ein furchtbares Blutbad angerichtet. Pferde mit zerschmetterten Gliedern und Menschen mit verzerrten Zügen lagen auf den Chausseen. Todesangst auf den Gesichtern. In die Chausseegräben hatten sie sich geworfen, um unseren Kugeln zu entgehen, und trotzdem hat sie das Schicksal erreicht. Es war ein furchtbarer Anblick.

An jenem 28. August sollten wir auch das erste Mal merken, dass der Krieg auch etwas Schönes mit sich bringen kann. Unser Einzug in Allenstein, der Jubel der Bevölkerung, - wie jeder was er hatte, gab, um uns zu erfreuen: Essvorräte, Bier, Wein, Limonaden, Taschentücher und was man nur irgend brauchen konnte, wurde uns zugereicht, und man hatte hier nun wenigstens den Erfolg all der Anstrengungen gesehen. Es war aber auch sehr nötig, denn nach den enormen Anstrengungen bei der grossen Hitze dieser Tage war es Balsam für uns, einen greifbaren Erfolg zu spüren.

Am Abend noch verliessen wir Allenstein, um ungefähr

8 Kilometer davon entfernt Biwak zu beziehen. Schon am nächsten Morgen um 4 Uhr ging es wieder los, - in Gewaltmärschen; ein ununterbrochener Trab von ca. 2 Stunden, dann 5 Minuten Schritt und wieder ein Trab von 3/4 Stunden, und bald wussten wir, dass es nicht mehr lange dauern würde, dass wir ins Gefecht kommen würden. Als wir aus einem Wald herauskamen, sahen wir Mannschaften der Jägerbagage in der grössten Aufregung uns entgegenlaufen, - dem Jäger-Bataillon war die Munition ausgegangen und sie fürchteten, sich zurückziehen zu müssen und die grosse russische Bagage im Stich zu lassen, die von den Russen hartnäckig verteidigt wurde. Aber schon waren wir aufgefahren, und in kaum 10 Minuten war der Sieg unser. Was von den Russen noch wegkonnte, lief in rasender Eile, und die riesengrosse Bagage, hunderte und aber hunderte von Wagen, 400.000.-- Rubel in barem Gelde waren unser. Vorher waren wir noch an einem Gehöft vorbeigekommen, kurz bevor wir in Stellung gingen, bei Wutrinen, wo am Tage vorher unsere Infanterie die Russen, die gerade beim Abkochen waren, überfallen hatte. Da sah man die Russen noch mit den Kochgefässen und Schüsseln in den Händen sitzen, so wie sie die tötliche Kugel erreichte. Es war mit das schauerlichste Schlachtfeld, das ich im ganzen Kriege gesehen habe. -

Am Abend hatten wir Biwak, und da wurde uns die Nachricht überbracht, dass der Feind in vollem Umfange geschla-

gen sei, und wurde uns hierdurch der wirkliche Erfolg der Schlacht von Tannenberg vor Augen geführt.

Am 31. zogen wir wieder in Allenstein ein und erhielten dort Ruhe bis zum 2. September. Am 31. schrieb ich aus Allenstein an meine Frau und will diesen Brief bis auf das Persönliche hier festlegen: "Heute haben wir einen sogenannten Ruhetag, das heisst, es darf sich niemand fortrühren, da natürlich jeden Augenblick Alarm sein kann. Und diesen Tag will ich nun benutzen, um etwas ausführlicher zu werden, als in letzter Zeit in meinen Nachrichten. Du schriebst gestern sehr richtig, dass Du davon überzeugt bist, dass wenn ich nur ein wenig Zeit habe, ich Dir ausführlich schreibe. Wie Du ja aber wohl aus den Zeitungen gesehen hast, haben wir hier wirklich Grosses geleistet. Nicht nur während der drei Tage schwere Kämpfe, sondern auch die Anmärsche von 12 bis 16 Stunden am Tage mit immer kleinen Kämpfen der feindlichen Nachhut. Und ebenso noch nachher, wenn man glaubte, alles sei vorbei, und dann kleinere feindliche Trupps sich immer noch zeigten und aus Gehöften und Wäldern auf unsere Leute schossen, wobei noch so mancher Brave sein Leben lassen musste. Unser Sieg war ein vollkommener. Fünf feindliche Corps vollkommen geschlagen, gefangen genommen und ihrer Bagage, welche immense Reichtümer enthielt, beraubt. Es wurden unter anderem in einer Kiste 400.000.-- Rubel in Gold und Silber gefunden. Hunderte und aber hunderte von Wagen mit wichtigen militärischen Sachen und Zeichnungen fielen in unsere Hände und wurden abgeliefert. An allem anderen hielten sich Offiziere, wie auch Mannschaften schadlos. Ich habe mir zum Prinzip gemacht, niemals etwas zu nehmen und führe das strikt durch. (- Ich habe das auch gehalten, bis ich nachhause kam und habe dadurch nicht, wie so viele Andere, Siegestrophäen mit nachhause geschleppt. Trotzdem tut es mir nicht leid,und halte ich meinen damals im Anfange des Krieges eingenommenen Standpunkt durchaus für den richtigen. -) Ungezählte Gewehre und andere Waffen, Kleidungsstücke, ganze Einrichtungen, sogar Frauensachen wurden gefunden, ein Zeichen dafür, dass diese Hunde sich nicht scheuen, in so schwerer Zeit ihre Frauenzimmer mitzunehmen.

Einen kleinen Eindruck von dem, was ich in der Schlacht empfinde, wirst Du ja wohl aus meiner Karte vom 26. während der Schlacht erhalten haben. (Ich schrieb damals: "Ich schreibe Dir, während der Kampf tost. Wir sind in einer Schlacht, einer furchtbaren, blutigen. Es war das Furchtbarste, was ich erlebt, nicht zu beschreiben. Wir haben einen sehr grossen Sieg, aber teuer erkauft, usw.")

Es lässt sich aber nicht in Worte fassen, wie schrecklich es ist, und wie sehr zu wünschen ist, dass dies grausige Schlachten bald sein Ende erreicht haben möge. Es werden schon vorher Anforderungen an die Leute gestellt, die zu überwinden wirklich nur ein einwandfrei gesunder Mensch imstande ist.

28 Stunden ohne Essen mit 2 Stunden Schlaf ist gang und gäbe. Und wenn ich im Nichtschlafen nicht eine so grosse Routine hätte, würde mir die Sache sicher weit saurer werden, als so, wo es mir, Gott sei Dank, wirklich sehr gut geht. Was man von dem Angstgefühl beim Anmarsch in eine Schlacht soviel gesagt hat, kann ich nicht bestätigen. Die ersten Minuten in der Schlacht sind unangenehm, wenn beim Einfahren in die Feuerstellung so rechts und links Schrapnells und Granaten neben einem einschlagen. Aber auch das verliert sich bald, wenn der Kampf so schön im Gange ist. Wenn aber nachher die ersten Verwundeten aus der Schlacht gebracht werden und des Schreiens und Jammerns kein Ende ist, dann erst wird einem klar, ein wie grosses, grosses Unglück über uns hereingebrochen ist. Ich will jetzt nur einige Einzelheiten schildern; aber wenn ich ein alter Mann mit grauem Haar werden sollte, werde ich das nicht vergessen, und dieser 26. August, unser Hauptentscheidungstag, wird mir immer im Gedächtnis bleiben.

Aber auch Gutes erlebt man wieder. Am Abend der Schlacht, wir hatten uns gerade ins Biwak begeben (nebenbei möchte ich nur noch bemerken, dass es heute gerade 8 Tage her sind, dass ich zum letzten Mal meine Stiefel ausgezogen habe), kommt die Nachricht, dass Allenstein seit dem 26. von den Russen besetzt sei, die Stadt und Land als russisch proklamierten, und dass die Russen Zeitungen in Deutsch und Russisch verteilt hätten, die Macht Deutschlands sei gebrochen usw. Also nachts um 2 Uhr Abmarsch. Um 1 Uhr mittags waren wir ca. 7 Kilometer vor Allenstein, als die Nachricht kam, das Gros des Feindes, das dann am 29. vernichtend geschlagen wurde, sei aus Allenstein fort und nur noch ca. eine Brigade sei in "Russisch-" Allenstein. Der Flieger, der die Nachricht von unserem Anmarsch nach Allenstein bringen sollte, war am Morgen von unseren Maschinengewehren heruntergeschossen worden, und so konnte sich unsere Artillerie 2 Kilometer vor der Stadt aufstellen, während unsere Infanterie die Stadt stürmte. Als die zu Tode erschrockenen Russen aus der Stadt flüchten wollten, fingen wir nun an, zu schiessen und richteten ein furchtbares Blutbad unter ihnen an. Unser Einzug nachher in Allenstein war wunderbar ! "Wir haben unsere Preussen wieder !" so ging es durch die Menge, die sich vor Jubel nicht zu lassen wusste. Gar nicht endenwollendes Hurrah. Brot, Bier, Wein, Limonaden, Zigarren, Zigaretten usw. Arme und reiche Leute, alle gaben gern ihr Letztes her. So manches Mütterchen stand da mit gefalteten Händen und betete. Es war ein rührend würdiger Anblick, die beglückte Bürgerschaft zu sehen. Aber schon ging es weiter. Ein kurzes Biwak, ein anstrengender Marsch, und wieder fing das Blutbad an. Schon als wir ankamen, sahen wir Menschen, Pferde und Wagen in tollen Haufen liegen und unsere Jäger gerade im Begriff, zurückzugehen. Aber da kamen wir, und in 5 Minuten hatte die "Jäger-Batterie", wie die Jäger unsere 6. Batterie getauft haben, den Feind geschlagen.

Ein schwerverwundeter Russe lag an der Strasse, niemand kümmerte sich um ihn. Er musste schon lange so gelegen haben, als ich zu ihm kam, ich holte den Sanitäter und den Arzt. Schuss durch die Lunge. Der Arzt gab ihm auf meine Bitte Morphium und verband ihn. Ich liess ihm zu trinken geben. Während das Blut über seine Lippen quoll , streichelte er den Sanitäter und mir nickte er zu mit einem Blick, den ich nie vergessen werde, und machte dabei eine Handbewegung, als wenn er sagen wollte: Ich konnte für alles dieses nichts. - Ich weiss nicht, ob er am Leben bleiben wird, weiss nicht, ob er jemand zuhause hat, der sich um ihn grämt, einen weiss ich aber, dem die dicken Tränen über die Backen liefen, während um ihn herum alles jubelte, - während ich Dir schreibe, muss ich wieder sehr toll schlucken, um den Tränen zu sagen: Ich bin Soldat ! - - Das sins so Kleinigkeiten, die sich einem so einprägen. Aber das grosse Ziel im Auge behaltend, soll man eigentlich über alles hinwegsehen. usw." -

Ein Monat Krieg war nun vorbei, und nach dem ersten Misserfolg, den wir hatten, war überwältigend Grosses geleistet worden. Und weiter ging es nun ohne Rast und ohne Ruhe, immer vorwärts, vorwärts, vorwärts. Von Allenstein nach Wartenburg, dort mal wieder ein Tag Rast, ein leidliches Quartier, ein Bett, sage und schreibe Bett ! Von Wartenburg nach Lisettenhof, dort merkte ich das erste Mal wieder, dass ich Gallensteine habe. Von Lisettenhof nach Mathildenhof, von dort nach Rotkerpen und von da nach Dietrichsdorf und weiter nach Arnsdorf. Hinter Arnsdorf bei Gerdauen hatten die Russen sich in ganz feste Stellungen begeben. Ueber unsere Marschleistungen brauche ich nichts zu sagen, nur soll man sich mal die Karte ansehen, und dann weiss man, was geleistet worden ist.

Am 2. September waren wir aus Allenstein fortgezogen und am 8. waren wir in Gerdauen. Gerdauen liegt auf einem Berge, und dorthin hatte sich das Gros der geschlagenen Russen zurückgezogen. Die Feinde hatten kleinere Truppenmengen zurückgelassen, um uns am schnelleren Vorwärtskommen zu verhindern. Den Zug des Gros konnte man genau feststellen, denn wenn die Spitze des Gros in ein Dorf kam, steckten sie es sofort in Brand, damit der Nachschub wisse, dort und dort seien die Russen und dort und dorthin hätten sie ihnen zu folgen. In der Dämmerung kamen wir in die Nähe von Gerdauen und bezogen dort ganz feste Stellungen, weil es sicher war, dass es zu schweren Kämpfen kommen würde. Und wir kamen in ein Artilleriefeuer von unglaublicher Heftigkeit. Am 10. früh sollte der Sturm auf Gerdauen angesetzt werden, aber von den Russen war nichts mehr zu sehen. In Nacht und Nebel hatten sie Gerdauen verlassen und wir mussten hinterher. Wie sah aber Gerdauen aus ? ! - Wir waren vorher schon einmal durch die blühende Stadt marschiert und kamen jetzt an einen jämmerlichen Trümmerhaufen. Es war nichts mehr stehen geblieben. Ganz vereinzelte Häuser, sonst nur Mauern, und diese Mauern hatten die Russen noch dazu benutzt, um sich dahinter zu befestigen. Die Befestigungen, die sie sowohl in der Stadt, wie auch an den Chausseen aufgeworfen hatten, waren mustergiltig. Und ein Sturm auf Gerdauen wäre wohl ein furchtbares Blutbad gewesen, so befestigt war es. Von unserem unübertrefflichen Hindenburg war es aber so gedacht, dass unser 17. Corps Gerdauen von der anderen Seite angreifen sollte, und das hatten die Russen aus der Schlacht bei Tannenberg schon gelernt. Lieber laufen, als in den Wurschtkessel kommen ! - Also wie schon gesagt, sie waren weg. Und nun ging es weiter. Genau, wie ich in den Tagen zwischen Allenstein und Gerdauen nur Städte erwähnt habe, durch die wir gezogen sind, so auch jetzt. Ueber Neuhof ging es nach Friedrichswalde, dann nach Melchersdorf, dort hatten wir ganze 2 Stunden Quartier, dann ging es weiter, und wir bezogen abends ein Biwak. In Melchersdorf erhielt unser Hauptmann das Eiserne Kreuz. Wir alle waren unglaublich stolz darauf. Am nächsten Tage ging es weiter über Bokellen, wo die Bahnbrücke gesprengt war, nach Harpenthal, einem kleinen Nest kurz vor Insterburg. Dort kam es wieder zu einem etwas schwereren Gefecht. Während der ganzen Tage vorher, - das möchte ich erwähnen, - waren selbstverständlich immer kleinere Gefechte gewesen. Bei Har-

penthal war meine Batterie einer Kavallerie-Division zugeteilt worden, sodass wir ganz an der Spitze marschierten und mit vorwärts mussten bei einer Attacke, die auf eine Chaussee nahe Insterburg angesetzt war. Hinter der Chaussee hatte sich feindliche schwere Artillerie aufgestellt, die uns glänzend einsehen konnte, sodass es ein Wunder zu nennen ist, dass wir ohne Verluste dort weggekommen sind. Mein gutes, braves Pferd Intarsia, auf dem ich sass, erhielt einen schweren Schuss. Es hätte auch anders kommen können! - Es ist einer der Gedenktage für mich in diesem Kriege, der 11. September. Ich ritt das Pferd noch ungefähr 8 Tage, trotzdem es sich elend quälte, dann wurde es in ein besseres Jenseits hinüberbefördert. Ehre seinem Andenken !

Aber zwischen allem Schrecklichen kamen auch immer wieder schöne Stunden. Und die schönste von allen war unser Einzug in Insterburg. Nein, nicht die schönste von allen, - die schönste war mein Einzug in Berlin. Aber in meiner Erinnerung ist gerade der Einzug in Insterburg etwas so wunderbares, dass es mir sicher für alle Zeiten unvergesslich sein wird. Insterburg war 3 Wochen von Russen besetzt gewesen, wo der General Rennenkampf und der Grossfürst Nikolajewitsch ihr Hauptquartier aufgeschlagen hatten. Eine halbe Stunde, bevor wir in Insterburg einzogen, sind diese beiden Schweinehunde hinausgelaufen. Schade drum ! Es wäre eine würdige Beute gewesen. - Aber von dem Jubel und der Begeisterung, mit der wir in Insterburg empfangen wurden, kann man sich keine Vorstellung machen. Da ver-

blasst der vorher so rühmlich erwähnte Einzug in Allenstein vollkommen. Die Leute waren selig. Sie bewarfen uns mit Blumen, kaum konnten wir mit unseren Pferden und Geschützen hindurch. Das Hurrah-Geschrei und die Begeisterung der Menge war einfach unbeschreiblich schön, und als gegen 8 Uhr abends die Kirchenglocken anfingen, zu läuten, zum Zeichen, dass Insterburg wieder deutsch sei, da traten auch dem Rauhesten von uns die Tränen in die Augen. Am Abend war die Stadt illuminiert, die Leute wussten wirklich vor Glückseligkeit nicht, was sie uns antun sollten.

Wir hofften,nun wenigstens etwas Ruhe zu bekommen, aber das ist ja, wenn man sich das heute in Ruhe überlegt, auch unmöglich gewesen. Gegen wieviel Fronten muss gekämpft werden. - Wie nötig ist jede kleinste Formation, und wenn wir draussen auch so manches Mal geflucht haben und gemeint, es ginge nicht weiter, so musste es doch immer wieder weiter gehen, denn es war einfach unmöglich, den Truppen Ruhe zu gönnen. Interessant ist es aber, auf der Karte nachzusehen, welch ein Weg vom 2. bis zum 11. September zurückgelegt worden war, und wenn man dabei noch besonders in Betracht zieht, dass wir für die Schlacht bei Gerdauen doch 2 Tage gebraucht haben.

Am nächsten Morgen ging es wieder weiter zur Verfolgung des Feindes. In der Nähe von Malwischken Biwak. Das schöne Wetter hatte uns verlassen, es regnete furchtbar. In diesen letzten Tagen hatten wir wieder 30.000 Gefangene gemacht und viel Bagage erbeutet. Bei schönem Wetter lässt es sich gut marschieren, das schlimmste ist Regen, viel schlimmer noch, als Schnee, den wir später auch kennen lernen sollten. Und dazu kommt nun die unglaubliche Müdigkeit. Ich hatte in den letzten 72 Stunden kaum 5 Stunden geschlafen. Jetzt in Inglauden, wohin wir am 13. kamen, hatte ich den ersten ausführlichen Schlaf seit Allenstein, denn wir hatten bis zum 14. mittags 2 Uhr Ruhe. Am 14. nachmittags 5 Uhr 7 Minuten wurde die russische Grenze bei Slowiki überschritten. Jetzt lernten wir erst die guten deutschen Wege schätzen. Wir kamen in einen Morast, der jeder Beschreibung spottet. Bis an die Knie sanken unsere Pferde ein, und vorwärts müssen wir, denn der Kanonendonner vor uns zeigte an, dass wir nicht weit vom Feinde waren. Um 7 Uhr waren wir wieder in schwerem Gefecht, und um 10 Uhr bezogen wir bei eisiger Kälte im Regen Biwak. Mir war noch am Abend mein grosser erster Vorratswagen in den Chausseegraben, soweit man hier überhaupt von einer Chaussee sprechen konnte, gefallen und mit Hilfe von Bäumen, die wir erst fällen mussten und mit enormen Anstrengungen haben wir das Ding wieder herausgeholt. Ich kam dadurch erst spät zur Batterie und wurde dort mit grosser Freude begrüsst, da man mich schon allgemein für tot gehalten hatte. - Schön war die Nacht nicht, es war die erste kalte Nacht. Ich schrieb in mein Tagebuch von eisiger Kälte. Was aber eisige Kälte ist, sollte ich erst später erfahren, in den schönen November- und Dezembernächten, wenn wir im Biwak lagen, und trotzdem glaube ich, dass wir noch keine Ahnung von dem haben, was unsere Truppen im Februar in den Karpathen in Bezug auf Kälte erlitten haben.

Am nächsten Morgen waren die sich immer so tapfer zurückziehenden Russen auch verschwunden und waren, - und das hat uns am meisten gekränkt, - ohne Munition mit 50 Geschützen über die Memel gegangen. Die 50 Geschütze hätten uns, glaube ich, noch gut gekleidet, aber es war nun vorbei. Weiter auf den schrecklichen Wegen ging es nach Gielgudiczky. Am 16. und 17. hatten wir Ruhetage. Das Gut, auf dem wir in einer Scheune lagen, (ich hatte die Scheune dem Hause vorgezogen und war klug gewesen, denn einzelne, die sich in das Haus gelegt hatten, waren zu unangenehmem Besuch gekommen) lag direkt an der Memel, und es war ein wunderbarer Blick vom Gute auf das Memeltal.

Am 18. ging es wieder zurück nach Deutschland. 13 Stunden ohne einmal Halt bei fürchterlichsten Wegen, bei entsetzlichem Sturm. Es war einer der schlimmsten Tage, die wir mitgemacht hatten. Am Abend kamen wir wenigstens in eine warme Wohnung zu deutschen Leuten, die uns freundlich aufnahmen. Die Stiefel aber durften wir nicht ausziehen, denn es wäre einfach unmöglich gewesen, sie am nächsten Morgen wieder anzubekommen, so waren sie vom Wasser getränkt. Schon am nächsten Morgen ging es weiter über das fast völlig zerstörte Eydtkuhnen nach Wirballen. Immer noch strömender Regen, aber wenigstens bis Eydtkuhnen deutsche Chausseen. In Wirballen führten wir ein wirklich gutes Leben. Wir kochten und brieten und entschädigten uns für all die Entbehrungen und Qualen, die wir durchgemacht hatten.

Meinen alten Wachtmeister Martin, mit dem ich mich so angefreundet hatte, verloren wir hier leider durch Krankheit und ist er auch nicht wieder zu uns zurückgekehrt. Mein neuer Wachtmeister, Herr Wachtmeister Schlieter, wurde sein würdiger Nachfolger, in jeder Beziehung würdig.

Ich füge jetzt wieder einiges aus meinem Briefe vom 21. September aus Wirballen ein: "....... Was wir hier durchgemacht haben, das lässt sich kaum in Worte fassen. Unser Oberbefehlshaber hat folgenden Armeebefehl erlassen, der die Vorrede zu dem sein soll, was ich Dir nachher schildern will. Armeebefehl vom 15. September 1914: Soldaten der 8. Armee ! Ihr habt neue Lorbeeren um Eure Fahnen gewunden. In zweitägiger Schlacht und in mehrtägiger rücksichtsloser Verfolgung durch Littauen hindurch bis weit über die russische Grenze hinaus habt Ihr die neue feindliche Armee, die aus dem 2., 3., 4., 20., 22. Armee-Corps, dem 7. Armee-Corps, der 1., 5. Schützenbrigade, der 53., 54., 56., 57., 72., 76. Reserve-Division, der 1., 2. Garde-Kavallerie-Division und der 1., 2., 3. Kavallerie-Division bestehende Wilna-Armee nicht nur geschlagen, sondern zerschmettert. Bis jetzt sind mehrere Fahnen, 30.000 unverwundete Gefangene, mindestens 150 Geschütze, viele Maschinengewehre und Munitionskolonnen, sowie zahllose Kriegsfahrzeuge auf den weiten Gefechtsfeldern aufgebracht worden. Die Zahl der Kriegsbeute nimmt immer noch zu. Eurer Kampfesfreudigkeit, Euren bewunderungswürdigen Marschleistungen, Eurer glänzenden Tapferkeit ist es zu danken. Gebet Gott die Ehre, er wird auch ferner mit uns sein. Es lebe Seine Majestät der Kaiser und König.

Der Oberbefehlshaber

von H i n d e n b u r g , Generalmajor.

Ich habe die "bewunderungswürdigen Marschleistungen" unterstrichen, weil gerade diese Marschleistung das Unmenschlichste war, was wohl je in einem Kriege geleistet ist. (Kurz nach Wirballen ist es viel schlimmer gekommen.-) Zwei Tage in Russland. Wege mit metertiefen Löchern. Täglich 18 Stunden im Sattel,ohne einmal abzusitzen. Dazu Regen und Sturm, dass man dachte, man wird mit dem Pferde umgeworfen. Uns allen, die wir doch an schwere Strapazen gewöhnt waren, sind zeitweise vor Ueberanstrengung die Tränen gekommen, zumal man Menschen und Pferde in Menge sah, die durch die Strapazen halbtot waren.

Manches Mal dachte ich, jetzt kann ich nicht mehr. Das Wasser lief mir am Rücken herunter. Trotz des Mantels, der über meine Stiefel gedeckt war, stand das Wasser in den Stiefeln, und der Regen, Hagel und Sturm peitschten in mein Gesicht. Aber der Mensch kann mehr, als er glaubt. - Und ich habe es geschafft, und wunderbarer Weise - die halbe Batterie ist schwer erkältet und ich nicht, obgleich ich die nassen Stiefel nicht von den Füssen bekommen habe.

Jetzt sind wir in Wirballen, einer echt russischen Stadt, in echt russischem Dreck. Aber wir haben Ruhe, wohlverdiente Ruhe. Manchmal habe ich gedacht, welche schweren Sünden ich wohl begangen haben muss, für die ich so habe büssen müssen. Aber ich habe gebüsst. Man hat in diesen Tagen seinen Gott kennen gelernt, der es wirklich gut, sehr gut meinen muss, wenn er einen das hat überstehen lassen. - Ich habe aber wenigstens immer gegessen, Brot und Butter mit Käse, die mir mein hervorragender Sell immer aufs Pferd gegeben hat. "Herr Unteroffizier müssen essen !" Wenn wir ins Quartier kommen, macht er mir Tee und, wenn möglich, Eier und bedient mich in geradezu mustergiltiger Weise. Um den Gefreiten Sell werde ich ebenso bei der ganzen Abteilung beneidet, wie um meine Post. Während ich hier schreibe, ist er in der Küche damit beschäftigt, mir Rumpsteak und Kartoffeln zu machen, wozu sich unser Wachtmeister, ein sehr netter Kerl, dessen gant besonderer Freund ich bin, und einige Kameraden eingeladen haben. Er hat mir ein ganzes Haus, das ich bewohne, das von seinen Bewohnern verlassen und von Russen und Deutschen geplündert worden war, wohnlich eingerichtet,sodass ich auf einem, wenn auch sehr harten und heruntergedrückten, Sofa liege, das so dreckig ist, dass wir eine Pferdedecke darüber gebreitet haben, mit einer zweiten decke ich mich zu. Dann wird ein Feuerchen angemacht, ein Tässchen Tee gebraut und man vergisst alles Schwere, das man durchgemacht hat und lebt wie ein Gent." - -

In der Nähe von Wirballen machen wir Schanzarbeiten, die, wie wir später merken, sehr notwendig gewesen sind. Am 22. kommt meine Beförderung zum Wachtmeister heraus. Ich bin der erste in der Abteilung, der überhaupt befördert wird.

Das Wetter war bis jetzt herrlich gewesen, aber nach den Erfahrungen, die wir schon aus der vorhergehenden Zeit hatten, konnten wir es uns denken, dass an dem Tage, wo wir wohl abrücken würden, sich auch das Wetter ändern würde. Und als am 27. Alarm kam, war der bis jetzt so schöne blaue Himmel trübe und dicke, schwere Regenwolken hingen vom Himmel. Nicht lange liess der Regen auf sich warten, und als wir am Abend von Wirballen in Mariampol eintrafen, waren wir schon bis auf die Haut durchnässt. Die Chausse war selbst für unsere Ansprüche ganz leidlich, sodass wir wenigstens schnell vorwärts kamen. Man munkelte, dass sich in der Gegend von Augustow starke feindliche Truppen zusammengezogen hätten, die wohl die rühmliche Absicht hatten, nach Ostpreussen einzudringen, und daran sollten sie gehindert werden. Die Märsche waren auch dementsprechend. Am 28. sollten wir von Mariampol bis nach Suwalki gehen. Es war aber einfach unmöglich, denn Regen und Sturm wüteten, wie wirklich noch nie im ganzen Feldzuge. Es wurde deswegen nachmittags gegen 6 Uhr in Andrejewo, einem kleinen Nest, wenn es überhaupt diese Bezeichnung verdient, denn es war noch kleiner, Halt gemacht. In einem Raume, wo eine sehr kinderreiche Familie lebte, wurden 40 Mann von uns untergebracht."Durchnässt" ist für unseren damaligen Zustand gar kein Ausdruck. Unsere armen Pferde mussten biwakieren und auch ein Teil unserer Mannschaften bei dem eisig kalten Regen, und manch einer, der in den nächsten Wochen wegen Rheumatismus usw. fortgekommen ist, hat den Grund zu seiner Krankheit an diesem Tage gelegt.

Am nächsten Morgen ging es über Suwalki nach Raczki. Ich schrieb in mein Tagebuch: Die Anstrengungen nehmen noch zu; Regen und peitschender Sturm, wie nie erlebt. - Es war einfach unmenschlich. Auch nicht eine Sekunde am Tage hörte der Regen auf, und immer schlug er uns von vorn entgegen, und wir gegen den Sturm ankämpfend. In Raczki hatten wir wenigstens gutes Quartier. Ich konnte auf einem Sofa liegen, mich ausziehen, bekam warmen Tee und ich sage es ganz offen, beinahe waren die Anstrengungen des vorhergehenden Tages wieder vergessen. Was kann der Mensch doch leisten, wenn er muss. - Lange konnten wir uns des schönen Quartiers nicht erfreuen, denn schon am 30. ganz früh ging es von Raczki nach Janowka und dort in Feuerstellung. Hier warteten wir auf den Feind, der nicht kam, und noch am Mittag änderten wir unsere Stellung und suchten uns eine neue ganz unweit von Wielgi-Pruski. Nachts hatten wir Ruhe, schliefen in einer Bude mit den kranken Bewohnern zusammen, die vor uns Barbaren, die wir ihnen bares Geld und zu essen gaben, eine höllische Angst hatten. Erst als sie merkten, dass wir wirklich nicht so schlimm seien, wie wir von den Russen geschildert worden waren, wurden sie zutraulicher. Am nächsten Morgen ging das Geballere los. Schon am Vormittag hatten wir zwei Verwundete. - Das Wetter blieb, wie es die Tage vorher gewesen war, und so schlecht wie das Wetter, so schlecht war der Erfolg.

Am 2. Oktober sollten wir eine neue Stellung einnehmen, die Russen hatten aber genau gemerkt, wie wir aufprotzen wollten, und uns mächtig ins Feuer genommen, sodass wir eilig zurückmussten, und mit heiler Haut kamen wir wieder an unseren alten Platz zurück. Am Abend kamen wir wieder in das Quartier bei Wielgi-Pruski, wo ein Arzt, der dort eingezogen war, sehr wenig damit einverstanden war, dass ich mein müdes Haupt bei ihm und seinen Kranken zur Ruhe legen wollte. Nach längerem Ueberreden entschloss er sich, mir die Erlaubnis zu geben. Wir haben uns sehr angefreundet. Es stellte sich bald heraus, dass er ein Freund von Dr. Domnauer und Dr. Meyersohn sei, und wir haben sehr nette Reminiszenzen gefeiert. Heute befindet er sich leider in russischer Gefangenschaft, in die er in den schweren Kämpfen bei Prassnitz im Januar geriet, jene Kämpfe, die meiner Batterie ja auch so viel brave Leute gekostet haben. Ich war im Januar aber schon nicht mehr dabei. - Wir assen und tranken zusammen, was jeder hatte, und ich schlief zwischen seinen Verwundeten. Als ich am nächsten Morgen erwachte, war der Mann, der fast neben mir lag, tot. Die Not bringt einen zu seltsamen Schlafgenossen ! -

Wie das Gefecht stand, wusste keiner von uns. Bald hörte man gut, bald weniger gut. Am 3. Oktober wussten wir es ganz genau: wir sollten uns neu gruppieren.- - Was das bedeutet, weiss jeder von uns. Eiliger Rückzug ! Unser linker Flügel hatte zwar gesiegt, aber dem rechten Flügel, an dem wir standen, war es dafür auch um so schlechter gegangen. Der Führer unseres Corps, General der Artillerie von Schubert, hörte damals auf, unser Corps zu führen. Es wurde eben "neu gruppiert."

Ich will bei dieser Gelegenheit, wenn ich auch bei manchem, der diese Aufzeichnungen vielleicht einmal lesen wird, auf Widerspruch stosse, erwähnen, wie wenig rühmlich sich unsere Kavallerie in allen den Kämpfen von Anfang an gezeigt hat, aber ich will denen sogar etwas entgegenkommen, die gegenteiliger Meinung sind, - ich behaupte es von der Kavallerie wenigstens, die unserem 1. Reserve-Corps zugeteilt war. Niemals hat sich das deutlicher gezeigt, als bei dem Rückzug, den wir jetzt anzutreten hatten.

Das Wetter war wieder schön geworden, ein Zeichen dafür, dass wir bald etwas Ruhe bekommen würden, und am 3. Oktober nachmittags kamen wir in Niedzwietzken an und sollten, da die Russen garnicht daran dachten, uns zu verfolgen, (wie von unserer Division versichert wurde, da sie selbst viel zu schwere Verluste erlitten hätten,) Ruhe bekommen. Ich wollte am 4. vormittags nach dem ungefähr 7 Kilometer entfernt gelegenen Marggrabowa fahren, als plötzlich in unserem Geschützpark Geschosse landeten. Unsere Kavallerie hatte so glänzend aufgeklärt, dass in dem Quartier, wo die Division lag, Kosaken waren. Ich habe hier das erste Mal erlebt, wie Preussen fluchtartig zurückgehen. Unsere grosse Bagage, die herangezogen war, stürmte die Chaussee herunter, wie die wilde Jagd, und wenn die Russen etwas klüger vorgegangen wären und nur eine halbe Stunde mit der Eröffnung des Feuers gewartet hätten, - was wäre wohl aus uns geworden ? ! - Wir gingen also in Feuerstellung, und nach 1 ½ Stunden hatten wir die Russen tatsächlich zurückgeschlagen. Ich schrieb in mein Tagebuch: Gefährlichste Lage im ganzen Kriege bis jetzt. - Es ist aber gut, dass ich hier gleich geschrieben habe: bis jetzt, - denn es sollte später doch noch ganz anders kommen. Ich wurde hier zum Führer der Staffel ernannt.

Am Abend verliessen wir Niedzwietzken, und in einer wunderbar warmen Herbstnacht zogen wir über Marggrabowa, das damals noch vollkommen stand, nach Schareyken, wo wir nachts um 12 Uhr anlangten. Hier kamen wir zu netten Leuten ins Quartier. Ihnen ist nachher auch alles genommen worden. Die Tochter, die nach Berlin reisen wollte, bat ich, meiner Frau Grüsse zu bringen, was sie auch getreulich getan hat. Ich glaube sogar, es hat ihr nicht einmal leid getan.

Schon am nächsten Morgen ½ 5 Uhr ging es weiter nach Goldap. Dort bestiegen wir die Eisenbahn, in Wagen untergebracht, die unserer würdig waren, nämlich in Viehwagen,denn viel mehr fühlte man sich ja doch nicht nach all den Strapazen und gar keiner Ruhe, und fuhren nach Eydtkuhnen. Abends kamen wir in Absteinen ins Quartier. Blutrot war der Himmel, - ein grosses Gefecht war hier im Gange. Die ganze Nacht wurde schwer geschossen; wir waren nur als Reserve aufgestellt. Ich hatte vorher erwähnt, dass wir bei Wirballen Stellung ausgeworfen hatten; diese Stellung nun hatte das Corps, das in so schwerem Kampfe lag, eingenommen, sie waren dort fast uneinnehmbar. Am nächsten Morgen um 5 Uhr zogen wir über Stallupönen auf die rechte Seite unserer Stellung, da der Feind versuchte, uns zu umgehen. Ihm mag es wohl unterwegs leid geworden sein, denn am Abend kamen wir zurück nach Eydtkuhnen ins Quartier, und dort blieben wir auch den 7., immer alarmbereit. Ein Waggon, der mit Wein und ähnlichen schönen Dingen angekommen war, brachte uns wenigstens für einen Tag in etwas freundlichere Stimmung. Ich schrieb in mein Tagebuch: Allgemeine Besäuftheit ! -

Am 8. Alarm. Ueber Bilderweilschen, Barzkehmen nach Kosakweilschen und dort Biwak. Am nächsten Tage, den 9. Oktober, Gefecht bei Schirwindt. Sieg auf der ganzen Linie ! Unsere Protzen wurden einmal vom Feinde derart beschossen, dass wir mehrere Pferde verloren und ein Unteroffizier verwundet wurde. Nachts kamen wir ins Quartier in ganz zugige Scheunen. Wunderbar schaurig war das Bild der brennenden Neustadter Scheunen, wo unsere Infanterie stürmend hindurchzog. Unsere Batterie stand auf einem Berg und unterstützte das Feuer, und es war wirklich grausig und erhebend zugleich, zu sehen, wie unsere braven Kerls da vorgingen.

Am 10. Oktober gingen wir wieder zurück nach Eydtkuhnen. Ich füge hier meinen Brief vom 10. Oktober ein: "Wo ich anfangen sollte, wollte ich Dir alles schreiben, was ich so in den letzten 14 Tagen, seit Du mein letztes ausführliches Schreiben erhieltest, erlebt und durchgemacht habe, weiss ich nicht, denn dann müsste ich Tag und Nacht schreiben. Und ich weiss, das willst Du nicht, denn ich bin totmüde. Wir haben seit dem Alarm, von dem ich Dir am 8. schrieb, wieder tüchtig im Feuer gestanden und den Feind, der uns gegenüberstand, völlig geschlagen bei Schirwindt. Aber es ist nur ein Teil der Schlachtenlinie,der Schlacht bei Wirballen, die, wie ich Dir ja schon schrieb, sicher eine der grössten dieses Krieges wird. (- Was sich nicht im Entferntsten bestätigt hat.-) Auf der linken Flanke des Feindes und besonders im Zentrum kämpfen unsere braven Truppen seit 7 Tagen und wenn, was wir nicht hoffen wollen, irgend etwas nicht gut geht, dann gibt es für uns schnell wieder Alarm und fort geht es ! - Darum ist von Ruhe auch jetzt noch keine Rede, man sitzt immer auf dem Pulverfass.

Doch nun mal in kurzen Worten, was sich so in den letzten 14 Tagen ereignet, was mir als bemerkenswert im Gedächtnis geblieben ist, denn durch die abwechselungsreichen Bilder verwischt es sich leicht. Das eine, was mir als schrecklichstes der Schrecken der letzten 14 Tage im Gedächtnis geblieben ist, ist das Wetter, das wir vom 27. September bis zum 3. Oktober hatten. Wie ich es beschreiben soll, weiss ich nicht. Also der Regen und Wind setzten an Heftigkeit all ihre Kraft ein. In meine Stiefelschäfte lief das Wasser durch meinen völlig durchnässten Mantel, sodass ich im Wasser in den Stiefeln stand. Auf dem Pferde konnte man es nicht lange aushalten, denn der Sturm blies einen fast herunter. Die ganzen Tage nicht die Stiefeln aus, aber vorwärts ging es, 16 bis 17 Stunden, nur mit der angenehmen Unterbrechnung der Gefechte, abends in einer kleinen Kammer zusammen 40 Mann - wer nicht wollte, durfte biwakieren -

um dann noch alles durchzumachen, was man nur erleben kann. Ich hatte aber diesen Stolz nicht, und so zogen wir lieber 40 Mann mit triefenden Kleidern, fast totgefroren, zusammen, um mit dem ersten Morgengrauen wieder auf das Pferd zu steigen und die Strapazen wenn möglich noch zu überbieten. Und es gelang ! Der schlimmste Tag war der 29. September, als wir von Andrejewo über Suwalki nach Raczki zogen. Das war der Anstrengungen schlimmste. Der Abend war aber gut. Ich hatte ein Jubiläum, wie mein treuer Sell ausgerechnet hatte. Genau vor 4 Wochen in Wartenburg hatte ich zum letzten Mal ein Bett gehabt, und dort winkte mir wieder eins. Von 9 bis 4 Uhr im Bett. Erst wusste ich nicht, was ich mit dem Deckbett anfangen sollte, aber Wonneschauer durchrieseln noch jetzt meinen Körper im Gedanken an ein richtiges Bett. Das war das einzige Mal in den Tagen, dass ich etwas von meinen Sachen ausgezogen habe. Den nächsten Morgen ging es ins Gefecht. - Es wurde heftig gekämpft, zwei arme Kerls meiner Batterie wurden verwundet. Auch den darauffolgenden Tag waren wir heftig im Feuer, und als wir am Abend einen Stellungswechsel vornehmen wollten, bekamen wir derartig heftiges Feuer, dass wir in unserer Stellung bleiben mussten. - Im Quartier, soweit man die russischen Dreckbuden so nennen konnte, traf ich einen Arzt, einen Vereinsbruder von Stoff und Perkeo, einen Dr. Schüler. Die Freude, mit jemand über Bekannte sprechen zu können, kannst Du Dir *ja nicht denken. Er war erst nicht sehr erfreut, dass ich in sein Quartier drang, dann haben wir uns aber so angefreundet, dass wir uns immer mit einander freuen, wenn wir uns sehen. Auch am nächsten Abend, der Kampf dauerte noch fort, waren wir zusammen. In meinem Zimmer lagen stöhnende Verwundete, am Morgen war einer tot. - - -*

Am nächsten Morgen zogen wir, vom Feinde unbemerkt, ab. Unser linker Flügel hatte gesiegt, über unsern Flügel schweigt des Sängers Höflichkeit. Am nächsten Tage, einem Sonntage, (es war der 4. Oktober), war Ruhe angesetzt worden. Ich lag und schlief in einer warmen Stube und freute mich, gerade an einem Ruhetage ein so schönes Quartier zu haben. Ich stand am Morgen auf, wusch mich (ich bemerke das besonders, da es auch nur zu den Seltenheiten gehört,) da kam plötzlich die Nachricht, dass uns der "geschlagene" Feind verfolge und dicht bei Niedzwietzken, so hiess das Nest, sei. Wir hatten die Nachricht noch nicht bekommen, da kam die erste Granate direkt neben unseren Geschützpark geflogen. Da habe ich zum ersten und hoffentlich auch zum letzten Male gesehen, was Flucht bedeutet. Unsere Bagage, dazwischen die armen Leute, die aus Furcht vor den Russen die Flucht ergriffen, - in heillosem Wirrwarr stürmten sie die Strasse entlang. Es war ein entsetzlicher Anblick ! Da kam aber eins zum Vorschein: die preussische Disziplin ! Unsere Leute waren mustergiltig. Die Granaten und Gewehrkugeln pfiffen um unsere Köpfe, aber alles war auf seinem Platze, direkt vor dem Feind. In 2 Stunden hatten wir ihn zurückgeworfen mit grossen Verlusten und vor allem mit Gottes Hilfe, denn wenn der Feind eine halbe Stunde länger gewartet hätte und uns nicht voreilig angegriffen hätte, - was nicht tot, wäre sicher in Gefangenschaft geraten. Alles ist aber der unglaublich funktionierenden Kavallerie zu danken, die im ganzen Feldzuge hier im Osten noch kein Ruhmesblatt in ihr Gedenkbuch geheftet hat. Der Feind hatte also seinen Sieg, ich bezeichne ihn wenigstens so, nicht ausgenutzt, denn wenn er grössere Truppenmengen uns nachgeschickt hätte, wäre das Ende wohl dunkel gewesen. Aber besser so ! Bei uns allen heisst es jetzt aber immer: Hoffentlich geht es uns nicht wieder so, wie bei Niedzwietzken. -

Ich will hier noch etwas erwähnen, und das ist der Name <u>Hindenburg</u>, der auf die Truppen wirkt, wie nichts anderes. Neulich ist erzählt worden, Hindenburg ist mit seinen Truppen im Rücken des Feindes, er wäre mit den Corps aus Lemberg abgezogen, um wieder zu uns zu stossen. (- Unsere irrige Ansicht war, dass Hindenburg mit seinen Corps bei Lemberg stand. -) Da waren die Leute wie aus dem Häuschen. Wie oft hört man sagen, wenn Hindenburg hier wäre, wäre dies und dies nicht so gemacht worden. Hindenburg wird bei unserer Ostarmee wie ein Heiliger verehrt.

Also am Abend dieses 4. Oktober marschierten wir wieder

los. Der Wettergott war uns gnädig und wir ritten durch eine wunderbare, nicht zu kalte, sternenklare Herbstnacht. Um 12 Uhr waren wir am Ziel in Schareyken, wo wir bis ½ 5 Uhr blieben. - Bei unserem Marsch durch Marggrabowa konnte ich noch ein Telegramm an Dich aufgeben, das hoffentlich in Deine Hände gelangte. Ich hatte es vom Pferde herunter einem Soldaten diktiert und einer Frau zur Weiterbeförderung übergeben. Das was das erste Mal, dass wir seit dem 27. September die Post trafen. Um ½ 5 Uhr morgens ging es weiter nach Goldap, und von dort wurden wir im Viehwagen nach Eydtkuhnen verladen, von wo ich Dir ja schon schrieb. usw." - - -

Aber nicht lange sollten wir uns der Ruhe erfreuen. Am 11. schon marschierte der Feind wieder auf Schirwindt zu, und weil wir unsere Sache damals so gut gemacht hatten, durften wir wieder hin. Schon bei Barzkehmen erhielten wir das erste Feuer, ein Zeichen, wie dicht der Feind uns auf dem Halse war. Ganz nahe am Feind, bei Barzkehmen, übernachteten wir. Nachts ½ 2 Uhr wurde eine russische Batterie gestürmt. Es war am 12. morgens ein schauriges Bild, diese gestürmte Batterie zu sehen. Pferde und Menschen lagen in wilden Haufen zwischen den Geschützen. Es wurde dort von einem russischen Leutnant erzählt, der aufgefordert wurde, sich zu ergeben und auf diese Aufforderung hin einem unserer deutschen Offiziers ins Gesicht schlug, und als er gefesselt wurde, sich losriss und nochmals dasselbe tat. Er wusste, was ihm die nächsten Minuten bringen würden. Es ist meiner Ansicht nach dies ein Zeichen ungeheuren Heldenmutes, etwas, was man in dem Feldzuge bei den Russen selten gefunden hat.

Ich habe schon öfters erzählt, und möchte das hier noch einmal niederlegen, wie abgestumpft man wird, denn gerade zwischen diesen Menschen- und Pferdeleichen wurde Halt gemacht und wir frühstückten. Dann ging es vor. In drei Stunden fünf Feuerstellungen. - Es war eine Hetze, wie nie zuvor. Ich dachte schon, dass man den Abend noch in Petersburg feiern wollte. Im ganzen waren an diesem Tage 30 Geschütze, 5 Maschinengewehre, 3000 Gefangene gemacht und viele, viele Tote lagen auf dem Schlachtfelde. Das Wetter war, solange wir in Deutschland waren, schön gewesen. An diesem Abend überschritten wir wieder die russische Grenze bei Schirwindt, und sofort fing der Regen an, sodass man bis zu den Knien in dem durchweichten Boden einsank. Ich hatte mich mit meinen Leuten in ein Haus begeben, um wenigstens einigermassen trocken zu sein. Den Leuten war durch ein Geschoss am Tage ihre Tochter getötet worden und der Jammer war entsetzlich. Trotz des Regens mussten wir am Abend wieder nach Deutschland hinüber. Wir hatten unsere Aufgabe auch diesmal wieder bei Schirwindt glänzend erfüllt und begaben uns wieder in unsere alten, zugigen Scheunen, wo wir nach dem Regenguss mörderlich froren. Am nächsten Morgen konnten wir in ein Haus gehen, uns einigermassen zu trocknen und dann kamen wir nach Barzkehmen zurück in ein wirklich gutes Quartier. Wir waren wieder in Deutschland, und sofort lachte der Himmel.

Am 14. waren wir wieder in Eydtkuhnen und blieben dort bis zum 18. Oktober. Das war eine schöne Zeit ! Am 16. fuhr ich nach Insterburg und konnte mit zu Hause telephonieren. Heute noch wird mir das Herz warm, wenn ich daran denke. Vom 3. August bis zum 16. Oktober niemand gesprochen zu haben und nun auf einmal die Stimme zu hören, das war herrlich ! Alle meine Erinnerungen an Insterburg sind schön. Als wir dort einzogen, als ich hier telephonieren konnte und noch eine, die ich aber erst später erwähnen will. Dankbar will ich hiermit der Damen vom Telephonamt in Insterburg gedenken, die, als ich ihnen meine Lage schilderte, dass ich seit Monaten meine Frau nicht gesprochen hätte, mir, obgleich es streng verboten war, gestatteten, zu telephonieren. Ich habe später ihre Liebenswürdigkeit auch noch einmal in Anspruch genommen und auch wieder mit demselben Erfolg, wie damals.

Am 18. Oktober gab es wieder Alarm. Auf unserem rechten Flügel mussten wir vor, weil ein Durchbruch der Russen zu befürchten war. Nachts biwakierten wir, und es war am 18. Oktober, wie man sich denken kann, sehr kalt. Zelte hatten wir auch nicht mehr. Am 19. warteten wir auf den Feind. Aber nichts war von ihm zu sehen. Wir sahen nur die Berge, von denen er kommen sollte, und freuten uns schon auf den Moment, wo wir ihm die ersten Grüsse hinüber senden sollten, denn wir hatten kolossal starke Stellungen bezogen. Es war aber nichts zu sehen, und am Mittag marschierten wir nach Dopönen, das vollkommen zerschossen war. Bei einem reichen Gutsbesitzer, der in einem Keller wohnte,. - das war alles, was ihm geblieben war - bezogen wir Quartier. Am nächsten Morgen sollte es weiter gehen. Die Geschütze waren ungefähr eine Viertelstunde von dem Quartier, wo ich mit meiner Staffel lag, entfernt, und als ich am nächsten Morgen mit meinen Leuten hinüberreiten sollte, verirrte ich mich und kam 1 ½ Stunde zu spät. Mir war nicht besonders wohl zu Mute, aber es ging sehr glimpflich ab. Ohne ein böses Wort wurde ich empfangen und schon nach kurzer Zeit zogen wir wieder in unser altes Quartier, da der Feind sich immer noch nicht rührte. Wir haben ihn auch an dieser Stelle nicht mehr getroffen. Später aber ist er doch noch über diese Berge gegangen und von den Truppen, die unsere Stellung dort bezogen, gebührend empfangen worden. Wer die Sache macht, ist ja auch ganz egal, - die Hauptsache, dass sie gemacht wird. -

Bis zum 26. Oktober blieben wir alarmbereit in Dopönen und lebten dort ein herrliches Leben, liessen uns Apfelkuchen backen, täglich wurde geschlachtet; die Frau des Gutsbesitzers kochte tadellos, und wenn wir auch in der Küche auf Stroh schliefen, auf das Betten gelegt waren, so war es doch eine gute Zeit. - Am 26. ging es nun wieder fort, von Dopönen nach Budweitschen. Dort hatten wir wieder Biwak, und am nächsten Tage ging es fort nach Bartnagora. Wir näherten uns wieder Augustowo, das uns noch nie Glück gebracht hatte. Am 28. gingen wir in Stellung bei Chmielowka und noch am Nachmittag hatten wir Stellungswechsel. Ich werde am Abend mit einem Zuge bis 200 Meter an den Feind geschickt, um bei dem Angriff, der erwartet wurde, unsere Infanterie besser unterstützen zu können. Zu wohl war mir nicht zu Mute, das kann ich nur sagen, denn wenn die Russen gestürmt hätten und unsere Infanterie hätte zurückmüssen, - Was wäre wohl aus meinen Geschützen geworden ? ! - Aber es ging alles gut, und ich war am nächsten Morgen wieder wohlbehalten bei meiner Batterie. Die Russen haben aber unsere Stellung zwischen Smolenska und Turowska genau eingesehen und haben solch ein Feuer auf uns eröffnet, dass wir auf Befehl alle unsere Geschütze verlassen mussten und in nahe Unterstände, die von der Infanterie schon vorher ausgeworfen waren, flüchten müssen. Die schwere feindliche Artillerie beschoss uns mit einer derartigen Heftigkeit, wie wirklich nie zuvor. Und ein Wunder ist es zu nennen, dass nur unser Beobachtungswagen, auf dem unser Hauptmann noch eine Minute vorher gestanden hatte, und unser 3. Munitionswagen, hinter dem ich kurz vorher gesessen hatte, ein Trümmerhaufen wurden. Ums Leben ist wunderbarer

Weise niemand gekommen. Nachher mussten wir an Stricken jedes Geschütz einzeln aus der Stellung ziehen, denn sobald wir uns den Geschützen näherten, feuerten die Russen wieder ganz fürchterlich. Leutnant Deuss wurde am Abend in den vordersten Schützengraben als Beobachtungsoffizier geschickt und machte dort den Sturm auf die Höhe 225 mit. Er ist dabei geblieben. - Wir haben ihm ein würdiges Denkmal unter einen hohen Tanne gesetzt. Ich will auf diesen Sturm auf Höhe 223 hier nicht näher eingehen und will auch das alles, was in den nächsten Tagen, dem 30. und 31. Oktober und dem 1. und 2. November passierte, übergehen, denn man ersieht am besten alles dies aus dem Briefe, den ich über dieses mein schwerstes Gefecht geschrieben habe:

"Vieles hat sich ereignet, seit Du die letzten ausführlichen Nachrichten von mir bekommen hast. Viele schwere Kämpfe, die selbstverständlich die nötigen Anstrengungen mit sich brachten, Kämpfe, deren Schwere alles bis jetzt Dagewesene in den Schatten stellt. Aber mit Gottes Hilfe habe ich alles gut überstanden, muss aber sagen, dass ich ähnlich noch nie im Feuer gestanden habe. Ich habe bei diesem Kampfe zum ersten Mal das Gefühl gehabt, dass mir etwas passieren könnte; denn sonst bekommt man eine kolossale Dickfelligkeit, wenn die Kugeln sausen, und ich hatte geglaubt, dass mich nicht aus meiner Ruhe während des Gefechts bringen könne. Hier bin ich doch mehrere Male in Situationen gekommen, wo ich kaum geglaubt habe, *heil herauszukommen. Aber die Erinnerung an alles Schwere ist etwas sehr Schönes, weil man es überstanden hat. Aber während man dabei ist, - ich sage es ganz offen, - überläuft es einen doch etwas kalt, wenn so 5 Meter entfernt die schweren Brummer neben einem einschlagen, oder auf einen Schützengraben, in dem man sich befindet, gerade das feindliche Feuer gerichtet ist. Aber alles will ich ja nachher noch ausführlich beschreiben.*

Das Wichtigste war, dass ich am Tage, wo ich in Dopönen mein wirklich gutes Alarm-Quartier verliess, noch den Pelz bekam, der herrlich ist, ebenso passt,und ohne den ich wohl unglaublich gefroren hätte, denn wir hatten bis 4 Grad Kälte, und mehrere Male war ich die ganze Nacht draussen, während ich die anderen Nächte in einer von uns selbst angefertigten Erdhöhle zugebracht habe, wo es ja bekanntlich nicht besonders warm sein soll. Und wenn wir auch so dicht bei einander schliefen, dass man sich nicht bewegen konnte, um wärmer zu liegen, wenn wir auch Mäntel und Decken hatten, so war es doch noch recht kalt. Ich will hier etwas ganz eigenartiges erzählen: Wenn ich im Freien schlafe, in der grössten Kälte, selbst in der Nässe liege, bin ich absolut nicht erkältet, sowie ich aber ein bis zwei Tage im warmen Zimmer war, bekomme ich den Schnupfen.

Also, wie gesagt, der Pelz und die Decken waren sehr, sehr angenehm, und haben sie mich zweifellos davor bewahrt, dass ich, wie schon mehrere, habe ausscheiden müssen, weil sie sehr schwer erkältet waren. Was kann aber doch der menschliche Körper ertragen ! - Man bedenke: bei 4 Grad Kälte Tag und Nacht im Freien. - Man gewöhnt sich aber an alles.

Nun will ich anfangen, wirklich zu erzählen, was war, was wir erlebt haben und wie uns zumute war, als wir in dunkler Nacht, vom Feinde unbemerkt, haben abziehen müssen. Es war, wie in den amtlichen Berichten stand, "Auf dem östlichen Kriegsschauplatz nichts neues". Wir haben aber bei Smolenska und Turowka, nahe bei Suwalki, wie mein Freund Schmidt sagt, (-wo er heute ist, weiss niemand. Hoffentlich in russischer Gefangenschaft; er ist eins der ersten Opfer von Prassnitz geworden,) "unanständig Dresche bezogen", allerdings gegen eine zehnfache Uebermacht, und sind abgezogen, um nicht noch grössere Verluste zu haben, denn der Feind hatte fast uneinnehmbare Stellungen eingenommen, gegen die anzustürmen ein heller Wahnsinn war, der blutig, leider sehr blutig bezahlt werden musste.

Also am 26. Oktober verliessen wir unser Alarm-Quartier, wo wir ein wirklich menschliches Dasein geführt hatten, denn bis auf das Wohnen in dem Keller und Schlafen auf dem Boden war die Sache dort sehr gemütlich. Das Essen war tadellos, und haben wir uns dort alle so recht erholt. Ueberhaupt ist das Essen und die ganze Verpflegung unserer Batterie hervorragend und ich glaube nicht, dass bei den Quanten, die ich einerseits aus *der Batterieverpflegung, andererseits und sogar besonders aus meiner eigenen Verpflegung vertilge, ich schon etwas abgenommen habe. (- ich hatte mich doch sehr getäuscht, - im Ganzen sind es 26 Pfund gewesen, die ich habe lassen müssen. -) Wir wussten, dass wir einen langen Marsch hatten, und nach einer sehr kurzen, sehr schlechten Nachtrast bei deutschen Leuten, die diesen Namen wirklich nicht verdienen - übrigens eine Erscheinung, die man des öfteren findet, dass die russischen Einwohner einen besser behandeln, wie die deutschen, die nur Angst haben, es könnte ihnen etwas fortkommen - zogen wir am nächsten Morgen wieder über die Grenze. Das milde Wetter, das wir in Dopönen hatten, war vorüber und es war recht kalt. Ueber Filipowo ging der Weg und von dort aus weiter. Nur, dass man den Weg nicht mehr als solchen bezeichnen konnte. Es war kaum etwas davon zu sehen, so der übliche russische Morast, sodass die Pferde manchesmal knietief im Dreck waten mussten. Aber der ferne und immer näher kommende Kanonendonner lehrte einen, dass es wieder anfing, ernst zu werden. Noch eine Nacht gab man uns Ruhe und am nächsten Morgen ging es in der Nähe des total zerstörten Chmielowka in Stellung. Schon beim Auffahren bekamen wir kolossales Gewehrfeuer, sodass unsere 4. Batterie schon beim Anfahren vor dem Absitzen den ersten Toten hatte. Unsere Batterie hatte eine schlechte Stellung erhalten und am Nachmittag schon mussten wir einen Stellungswechsel vornehmen, der uns sehr leicht hätte verhängnisvoll werden können, wie es sich am nächsten Tage herausstellte. Aber erst kam der Befehl, dass ein Zug (2 Geschütze) am Abend bei eintretender Dunkelheit nach vorn zur Infanterie musste, die bis auf 200 Meter vor dem Feinde war. Also mich betraute der Hauptmann mit diesem ebenso ehrenvollen, wie zweifelhaften Auftrag. Ich hatte noch nie einen Zug geführt, sollte mir meine Stellung allein aussuchen und sollte die Infanterie unterstützen, die einen feindlichen Sturmangriff erwartete, und sollte meine Geschütze und Leute gesund nachhause bringen. Ich habe gebetet: Lieber Gott, lass mich hier erst gut rauskommen ! - So unparlamentarisch es klingt: ich hatte die Hosen gestrichen voll !*

Die Russen sind grosse Schweine, und ich habe im ganzen Feldzuge nur einen einzigen guten Zug bei ihnen entdeckt: - sie liessen uns diese Nacht in Ruhe. - Wenn ich an diese Nacht zurückdenke, danke ich ihnen jetzt noch dafür. Beim Morgengrauen rückten wir ab und waren eine Stunde später wieder glücklich zurück, bei unserer Batterie. Nun dachten wir nach der durchwachten Nacht einige Stunden Ruhe zu erhalten, als plötzlich die schwere feindliche Artillerie anfing, uns zu beschiessen, und zwar ein Feuer, wie noch nie zuvor. Jeder Schuss lag in der Batterie und der Hauptmann gab den Befehl, die Geschütze zu verlassen und uns in die aus früheren Gefechten angelegten Unterstände für Infanterie zu begeben. Was es für einen preussischen Soldaten heisst, sein Geschütz zu verlassen, kann sich jeder, auch der Nicht-Soldat, vorstellen. Es war furchtbar ! Unser Beobachtungswagen,auf dem unser Hauptmann noch 5 Minuten vorher gestanden hatte, glich einem Trümmerhaufen. Der 3. Munitionswagen, hinter dem ich gestanden, war glatt durchlöchert. Und wenn die Hunde nur etwas besser geschossen hätten, wäre die ganze 6. Batterie in kurzer Zeit ein Vexierbild geworden. Nach einer Stunde wagten wir es, aus unseren Unterständen herauszugehen. Aber schon eröffnete der Feind das Feuer von neuem. Wieder ging es zurück. Nach langer Zeit mussten wir die Geschütze und Munitionswagen einzeln aus der Stellung herausholen, und wunderbarer Weise, Gott sei Dank, gelang es uns ohne Verluste, und eine neue Stellung nahm unsere heilen Wagen und Trümmer auf. Ich sage es aber, oft möchte ich nicht solche Stunden verleben, die gehen schwer auf die Nerven. (- Hier hatte ich meinen ersten wirklichen Gallenanfall bekommen. Ich hatte es aus leicht verständlichen Gründen in dem Briefe nicht erwähnt. -) Meine Feder ist leider nicht beredt genug, um alles in seiner unglaublichen Gefahr für unsere Batterie schildern zu können. Wenn ein Tag schlecht anfängt, dann endet er auch schlecht! - Am Abend sollte ein Offizier zur Beobachtung vorgeschickt werden, etwas, was ich -nachher noch näher erklären will. Unser Leutnant Deuss wurde dazu kommandiert. Bei dem schweren Gefecht musste er, von einem Infanteriegeschoss verwundet, sein Leben lassen. Seine Leiche haben wir unter einer herrlichen Tanne beigesetzt. Er fand seinen Tod bei einem Sturmangriff von uns gegen die für mich unvergessliche Höhe 223 bei Suwalki, die nach einem halbstündigen Kampfe von uns genommen wurde. Unser Reserve-Regiment 61 hat in dieser halben Stunde 900 Mann, - sage und schreibe 900 Mann verloren. Und der Erfolg ? - Nachts kam der Befehl, die unter so schweren Opfern erstürmte Höhe wieder zu räumen, ein Befehl, der uns allen rätselhaft geblieben ist, denn alle Verluste der nächsten Tage und unser ruhmloser Rückzug sind nur auf das Konto der Räumung dieser Höhe zu schreiben. Viele, viele Verwundete liegen noch dort oben, die nicht mehr haben fortgeschafft werden können, und sind in ihren Schmerzen teils verhungert, teils erfroren. Das war das schlimmste Bewusstsein, das uns die Tage hindurch beseelte: die armen Leute, die dort oben liegen und auf die zuhause gewartet wird. - Welch Unglück ist der Krieg, - und über alles geht man zur Tagesordnung über ! -

Ich habe in mein Tagebuch über den 30. Oktober geschrieben: Das schwere Gefecht dauert fort. Wir bauen uns eine Höhle. - Das war der Unterschlupf, von dem ich schon vorher berichtete. Die Leute haben die Höhle mit grossem Geschick gemacht und direkt neben der Kanone angelegt. Jede Kanone hat ihre Höhle, und überbieten sich die Leute natürlich, welches die schönste ist.

Sie ist ungefähr 1 Meter hoch und 4,50 Meter breit, ist oben mit Erde, Brettern, Laub und Stroh bepackt und auch innen an den Wänden und auf der Erde mit Stroh ausgelegt. Wenn die Leute am Tage frieren, arbeiten sie an der "Wohnung." Und wenn abends die Lichter innen brennen und 7 Leute drinnen liegen, kann es manchmal ganz mollig warm werden, überhaupt, wenn tüchtig innerlich mit "Petroleum", das ist nämlich Rum oder Cognak oder ähnlich erwärmende Mittel, eingeheizt ist.

Einer unserer Offiziere ist zur Beobachtung in den Schützengraben geschickt. Dort sind sämtliche von der Batterie abgegebenen Schüsse gut zu kontrollieren, da meistens das Ziel gut zu sehen ist. Mittels Telephon, das des Nachts gelegt ist, wird nun die Batterie verständigt, wie sie zu schiessen hat, und in welcher Weise und von welcher Seite unsere eigene Infanterie angegriffen wird. Nach 24 Stunden soll der Leutnant abgelöst werden. "Vizewachtmeister Riess, Sie können Leutnant Glaubke ablösen !" - - Bei hereinbrechender Dunkelheit mache ich mich auf den Weg, bekomme verhältnismässig wenig Feuer, sehe aber, was die grossen Brummer angerichtet haben, - überall metertiefe Löcher, die Häuser, die 2 Stunden vorher noch standen, sind fast vom Boden verschwunden. Hier liegt ein Gewehr, an dem man noch Stellen von dem Blut sieht, das sein armer Besitzer hat lassen müssen. Dort ein Tornister mit allen möglichen Sachen, sicher einem Verwundeten gehörend, der die schwere Last nicht hat bis zum Verbandplatz schleppen können. Und immer wieder kommt es einem zum Bewusstsein: welches Unglück ist der Krieg !

Also ich ziehe mit meinem Telephonisten los und erreiche nach gut einhalbstündigem Marsch den Schützengraben, lasse mich von dem Leutnant orientieren, was bei der Dunkelheit gut geht, sehe keine 50 Meter vor mir die furchtbare Höhe 223 mit allen ihren Schrecken und verlebe einen recht gemütlichen Abend bei den Offizieren der Maschinengewehr-Abteilung, die in dem Schützengraben lagen und sich eine herrliche Höhle gebaut hatten. Viele Lichter und noch mehr Menschen sorgten neben dem Seligmacher Cognak für die nötige Wärme, und da uns auch der Feind in Ruhe liess, war es sehr, sehr nett.

Ein sehr nettes Episödchen ist noch ganz ulkig: selbstverständlich in derselben Höhle, wie der Hauptmann und Oberleutnant, wohnen auch die Burschen, schon ältere Reserveleute, waschechte Berliner. Also der Hauptmannsbursche, Herr Wilhelm Meier aus Pankow, war eine Type. Im Gespräch erzählte ich den Offizieren, wer uns was ich bin, und als besagter Wilhelm Meier aus Pankow hört, dass ich von G. G. bin, stösst er ein Indianergeheul aus. "Also Herr Hauptmann, da kaufe ich alles, wir alle kaufen da. Ein so reelles Geschäft !" - Bis der Hauptmann ihm sagt: "Meier, Sie wollen doch nur bei dem Wachtmeister eine Zigarre erben, Sie alter Schmuser !" Als aber auch der andere Bursche ein Loblied auf G. G. anstimmte, war der Hauptmann besiegt und wir lachten wirklich herzlich, wie man es in der Situation sonst kaum tun würde. Berühmtheit von G. G., sogar im Schützengraben zwischen Smolenska und Turowka ! -

Um 1 Uhr nachts wurde das Regiment abgelöst von einem Bataillon. Ein Regiment war schon vorher mit 3 Batterien von uns fortgezogen,um dem Feind bei Schittkehmen erfolgreich entgegenzutreten. 2 Batterien nahm auch dies Regiment wieder mit und ein Drittel blieb, um denselben Platz auszufüllen, den noch 2 Tage vorher 2 Regimenter eingenommen hatten. Als die Truppen fortzogen, sollte der Schützengraben, in dem ich mich befand, neu besetzt werden. Aber nichts kam, und als um $\frac{1}{2}$ 4 Uhr auch jede telephonische Verbindung gestört war, musste ich, bevor es tagte, fortziehen und kam zur Batterie, meldete mich und wollte mich ein wenig zur Ruhe begeben, denn ich hatte ja seit 36 Stunden nicht geschlafen. Da musste ich zum Hauptmann, der mir mitteilte, ich müsste versuchen, in den Schützengraben zu gelangen, legte mir ans Herz, mich aber um Gottes Willen nicht in Gefahr zu begeben. Das waren nun sehr schöne Worte, aber eben so unausführbar. Aber ich wollte in den Schützengraben, unter allen Umständen. - Einen guten Teil des Weges konnte ich mich decken, obgleich immer neben uns, dem Telephonisten und mir, die Geschosse einschlugen. Als wir ungefähr 300 Meter vor dem Feinde waren, stellten wir uns hinter ein Haus und warteten eine Weile, bis es ruhig wurde, dann Sprung auf, marsch! - Nieder!-

wieder auf , wieder nieder, - und neben und über uns krachten die Geschosse, und als wir nun glücklich in den Schützengraben gelangt waren, ging das Feuer auf diesen Schützengraben los. Fast an die Wand gedrückt, stand ich da und beobachtete, wie wir und wie der Feind schoss. Wie schon gesagt, sollte der Graben frühmorgens von Infanterie besetzt werden, aber niemand war da, - die ganze Flanke war gedeckt durch einen Vizewachtmeister und einen Mann. Wenn das die Russen gewusst hätten ! - Ich liess dem Hauptmann wörtlich durch das Telephon sagen: wenn ein Flankenangriff erfolgen sollte, hätten wir als einzige Waffe unser Telephon, und damit würden wir schmeissen. - Es soll grosse Heiterkeit hervorgerufen haben.

Ich sollte noch die ganze Nacht im Schützengraben bleiben. Es kam um 3 Uhr nachmittags der Befehl, bei hereinbrechender Dunkelheit das Telephon abzubrechen und zurückzukommen. Ich war so glücklich, ich wusste vor Freude kaum, was ich machen sollte. Die alten Truppen hatten alles aus der Höhle herausgenommen, es war eisig kalt und ich musste die ganze Zeit im Freien stehen und beobachten. Sobald es dunkel wurde, zogen wir los, mit klappernden Zähnen, nichts im Magen seit 24 Stunden. Aber als ich zur Batterie kam, bekam ich ein dickes Lob. Als ich zum Hauptmann kam, waren sämtliche Offiziere versammelt. Der Hauptmann sagte mir: "Ich spreche Ihnen meine vollste Anerkennung aus für Ihr tadellos tapferes Benehmen. Sie haben Ihre Sache *ausgezeichnet gemacht." - Dann musste ich von meiner Expedition in den Schützengraben am Vormittag erzählen, und als ich geendet, sagte mir der Hauptmann, dass wir abziehen, zurück nach Deutschland in Nacht und Nebel. Ein gemischtes Gefühl ! Einerseits Deutschland, das so unglaublich lockte, andererseits der Rückzug, der etwas so deprimierendes hat. Aber ich war froh, dass ich aus der Sache mit heiler Haut herausgekommen war.*

Wir marschierten dann die ganze Nacht bis morgens 7 Uhr. Durch jedes Dorf, durch das wir kamen, immer der Jammer der Einwohner: Jetzt kommen die Russen wieder. - Und sie hatten Recht. - Der Feind, der nichts bemerkt hatte, feuerte wie toll auf unsere verlassene Stellung. Hoffentlich hat er tüchtig Munition verschossen. Die Kugeln haben aber wenigstens keinem wehe getan. Aber das Gefühl beschlich uns wieder: die armen Kerls da oben auf Höhe 223, die wie wir für ihr Vaterland gekämpft hatten und nun schnöde im Stich gelassen wurden. -

In der Nacht wäre ich vor Müdigkeit zweimal fast vom Pferde gefallen. 70 Stunden hatte ich keine Minute Ruhe gehabt und dabei 20 Stunden ununterbrochen scharf beobachten müssen. Ich war wie gerädert. Am Tage fand ich keine Ruhe, so übermüdet war ich, aber in der Nacht habe ich geschlafen wie ein Toter. Wir waren wieder raus aus Russland, das wir alle hassen, wie die Sünde. usw." - -

Schon früh am Morgen ging es aus Gurren fort. Deutsche

Chausseen und deutsches Wetter ! Man kommt sich vor, wie auf einer Felddienstübung. Ich hatte damals, als ich von Chmielowka herunterkam, nie das Gefühl gehabt, dass wir zurückgingen, sondern bei mir hatte immer das Gefühl die Oberhand behalten, dass ich glücklich aus dieser furchtbaren Lage dort oben herausgekommen war, aus der Lage, die mir, wie mir ja mein Hauptman einige Tage später sagte, das Eiserne Kreuz eintrug.

Von Gurren ging es nach Gross-Wischteken, und schon auf dem Wege nach dort wurde ich vom Hauptmann beauftragt, nach Insterburg zu fahren, und dort sollte ich die Nacht bleiben. Ich glaube, ich habe noch nie derartig gestrahlt. Ich konnte telephonieren, in einem Bett schlafen, baden, - das letztere das erste Mal im Feldzuge - und, und das war noch wichtiger, als alles andere, - ich konnte mit Zuhause sprechen. Ich sage nur wieder: Gesegnetes Insterburg ! Meine Telephondamen, die mich seinerzeit schon so liebenswürdig aufgenommen hatten, halfen mir auch diesmal wieder, und lange, ausführliche Gespräche mit Berlin und Stettin machten mich für einige Stunden zu einem glücklichen Menschen.

Es war schon alles Mögliche gemunkelt worden, dass wir verladen werden sollten. Keiner wusste, wohin, hunderterlei wurde gesagt. Die einen sagten: nach Westen, die zweiten sagten: nach der Türkei, einige sagten: zur Besatzung einer Festung, und einige: nach Warschau. Die letzteren hatten Recht, d.h. nach Warschau wurden wir nicht verladen, denn unsere Truppen, die kurz vor Warschau standen, hatten in Eilmärschen zurückmüssen, weil unsere treuen Bundesgenossen einmal wieder versagt hatten und unsere Truppen so den Anschluss mit den Oesterreichern verloren hatten, sodass es leicht zu einer bösen Katastrophe hätte führen können. Unser genialer Führer Hindenburg hatte sich aber seinen Rückzug so grossartig ausgearbeitet, dass er mit einem grossen Siege für uns endete. Und wenn wir auch heute noch nicht an den Stellen stehen, wo damals unsere Heere gestanden haben, so ist doch die unglaubliche Zahl von Kriegsbeute, Kriegsmaterial und Gefangene, ein gutes Aequivalent dafür.

Also in Insterburg lebte ich nun einen schönen Abend. Um das Mass der Genüsse voll zu machen, ging ich auf 5 Minuten sogar in einen Kientopp, aber länger als 5 Minuten hielt ich es auch nicht aus. -

Am nächsten Tage wurden wir verladen. Wohin, wussten wir immer noch nicht. Plötzlich hielten wir in Thorn. Unsere Geschütze und Pferde wurden ausgeladen, und weiter ging es nach Klein-Morin. Das war mal ein Quartier ! Die braven Leute, Herr Hammermeister und seine Frau, die wussten, was uns gut tat. Wir haben dort bis zum 10. gelegen, und ich habe einen Tag bei unseren lieben Freunden, Frombergs, in Thorn verbracht, habe mal wieder von jemand erzählen hören, wie es meinen Lieben zuhause ging und bin eben bei Leuten gewesen, die ein wirklich persönliches Interesse für einen hatten. Das war eine grosse, grosse Wohltat für mich, und noch einmal so leicht bin ich wieder zurück zu meiner Batterie gefahren.

Wir waren jetzt durch die letzten 5 Tage recht verwöhnt worden. Aber schon am 10. November, als wir von Klein-Morin nach Serovski gingen, merkten wir den Unterschied. In einer wüst stinkenden Bude brachten wir die Nacht zu, und jetzt ging eine Zeit los, die an Strapazen alles überbot. Vom 10. November bis zum Tage meiner Abreise haben wir keinen Tag Ruhe gehabt, und auch später, als ich schon fort war, hat es noch eine ganze Weile gedauert, bis meine Batterie einen Ruhetag erhielt. - Am 11. November schon stiessen wir auf den Feind, der uns bei Jaranowek starken Widerstand entgegenstellte. Auch den 12. wurde an verschiedenen Orten heftig gekämpft, aber immer ging es vorwärts. Das Gefecht dauert auch am 13. noch fort, und der Feind, der hier nicht allzustark zu sein schien, zieht sich fluchtartig zurück. Wir kommen abends ins Quartier nach Wieslawice, ein wunderbares Schloss mit einem Stall und Pferdematerial, wie ich es nie schöner gesehen habe. Ich glaube, am nächsten Morgen wird sich der Besitzer gewundert haben, wieviel weniger gut seine Pferde geworden waren, denn mancher brave Zosse, der den Feldzug von Anfang an mitgemacht hatte, wurde zur Ruhe nach Wieslawice eingestellt. Da wir aber unser Pferdematerial vollständig haben mussten, so wurde dafür ein anderes"eingetauscht."- -

Dass nun das Gut von aussen so glänzend aussah und auch innen von den Reichtümern des Besitzers zeugte, liess noch nicht ausschliessen, dass sich unsere halbe Batterie auf dem Gut und den Gehöften um das Gut Lause holte. Gegen mich müssen wohl die Lause eine Antipathie gehabt haben, denn ich habe im ganzen Feldzuge keine gehabt, was ich aber wohl besonders zwei Leuten zu danken habe, erstens meiner Frau, die mich reichlich mtt frischen Sachen bedachte und mir Insektenpulver in grossen Quantitäten zukommen liess, und meinem treuen Sell, der dafür sorgte, dass ich die frischen Sachen anzog, denn von selbst sind wohl die wenigsten von uns darauf gekommen, mal die Wäsche zu wechseln.

Von Wieslawice ging es über Kowal auf Gostynin los, wo sich 3 russische Corps befanden. Ein schwerer Kampf war für den 15. zu erwarten, der damit eingeleitet wurde, dass wir am 14. November ohne Zelte biwakierten. Wenn man am 14. November in Berlin ohne Mantel gehen würde, würde man für verrückt erklärt werden. Dort hat es uns oder wenigstens den meisten von uns aber nichts geschadet, dass wir - zum grössten Teil ohne Decken - dort die Nacht verhältnismässig sehr schön schliefen. - Der 15. November brachte uns wieder einen sehr schönen Erfolg. Aber schwer war der Kampf, und der Feind muss an

Toten Unglaubliches verloren haben, wir aber auch ! ! 6000 Gefangene, 6 Geschütze, darunter zwei schwere, und 15 Maschinengewehre fielen in diesen Tagen in unsere Hände. Man bedenke, unser eines Corps stand 3 russischen gegenüber und rieb sie so auf, dass sie sich fluchtartig zurückziehen mussten. Wir verfolgen den Feind über Bialtotar, wo man ein furchtbares Schlachtfeld sieht. Es ist eisig kalt, als wir nachts um 1 Uhr in Konstantinowo ankommen und, sage und schreibe, bei 7 Grad Kälte Biwak beziehen. Als ich am nächsten Morgen aufwachte, dachte ich, ich könnte kein Glied mehr rühren, es wäre mir alles erfroren. Aber es ging. Es geht eben alles, was man muss. Schon eine halbe Stunde später sassen wir, ohne etwas Heisses in den Magen bekommen zu haben, auf unseren Pferden. O Gott, wie sah das Schlachtfeld aus, an dem wir jetzt vorbeiritten ! Ich entsinne mich: eine russische Batterie hatte sich vor einem Gehölz aufgestellt und war von einer schweren Batterie von uns beschossen worden. Bald sah man an dem einen Baum einen Arm, bald ein Bein und bald einen Rumpf hängen, und zwischen den total zerschossenen Geschützen lagen die Pferde. Selbst für uns, die wir doch schon an viel Grausiges gewöhnt waren, war es entsetzlich. -

Wir verfolgen den Feind am 16. durch Gostynin, und bald hinter Gostynin stossen wir auf die Nachhut des Feindes, die uns starken Widerstand leistet. Nachts hatten wir Ortsunter-

kunft in Gatschno. Am 17. stossen wir bei Gudinowia auf den Feind, das heisst, wir waren noch in Marschkolonnen und garnicht auf einen Angriff vorbereitet, als wir heftiges Infanteriefeuer bekommen. Dort blieben wir den Tag und auch die Nacht. Am nächsten Morgen hatte sich die Nachhut des Feindes, denn diese war es natürlich immer nur, wieder zurückgezogen und wir gehen von Gombin nach Kamin. In Kamin wollten wir die Nacht bleiben und sollten am nächsten Morgen in Feuerstellung gehen. In der Nacht wurden wir plötzlich geweckt, - in einem von uns besetzten Hause war Feuer ausgebrochen, sicherlich von einem Russen angesteckt. Es war ein schauerlicher Anblick! Ein armer Kerl von uns hat sein Leben dabei lassen müssen. Furchtbar verbrannt kam er aus dem Feuer schreiend heraus, er ist kurze Zeit darauf unter furchtbaren Schmerzen gestorben.-Am nächsten Morgen gingen wir in Feuerstellung. Es war uns gesagt worden, dass der Feind, der dort kolossal stark wäre, vollkommen umzingelt sei und jetzt der grosse Sedan-Tag anbreche. Wir warten nun in Kamin, dass der arme Feind irgendwo, vielleicht auch auf unserer Front, versuchen wird, dem "unentrinnbaren" Schicksal zu entgehen. Wo war er aber ? - In letzter Minute hatte er einen Durchschlupf gefunden und war wieder aus dem Wurschtkessel heraus. Also hiess die Parole: Weitermarschieren ! Wir marschierten bis zum Abend durch schönes Land, auf schlechten Chausseen an teilweise herrlichen Gütern vorbei und wollen am Abend Quartier beziehen. Aber schon aus dem Dorfe vor unserem neuen Quartier erhalten wir starkes Infanteriefeuer, blieben noch ungefähr 2 Stunden im Gefecht und ziehen auf dem schönen Gut Luszyn ins Quartier. Am nächsten Morgen gehen wir bei Luszyn in Stellung. Ich befinde mich hinter dem 5. Munitionswagen, als ich sehe, dass mein 4. Geschütz falsche Richtung genommen hat. Trotz des Warnens des Geschützführers, mich nicht hinter dem Munitionswagen hervorzurühren, da der Feind uns ziemlich genau eingesehen hat, gehe ich zum 4. Geschütz hinüber. In derselben Minute wird die Bedienungsmannschaft, die hinter dem 5. Munitionswagen gestanden hatte, weggerafft. Dem einen ist der Kopf heruntergeschossen worden, von dem andern war überhaupt nichts mehr zu erkennen, - und da soll man nicht an Bestimmungen glauben ! -

Am Abend kamen wir nach Stempow ins Quartier. Von Stempow gehen wir nach Dlugi in Feuerstellung und von dort in eine neue nach Slawkow. Bei Dlugi sehen wir viele verbrannte Leichen liegen, - man kann mit ziemlicher Sicherheit annehmen, dass ein Verbandplatz von unserer Artillerie eingeschossen worden war. Ich erwähne das mit Willen, da so oft in den Zeitungen zu lesen ist, dass die Feinde Verbandplätze eingeschossen hätten. Es ist jedoch für Artillerie unmöglich, genau zu wissen, wohin man schiesst. Wir hatten damals den Auftrag bekommen, den Ort Dlugi unter Feuer zu nehmen, und diese Aufgabe war erfüllt worden. War es uns denn auf eine Entfernung von ca. 5000 Meter möglich, zu sehen, wohin wir schossen ? - Und so kann es eben vorkommen, dass, ohne dass man es will - und das setze ich bei unseren Feinden wie bei uns voraus, dass nicht mutwillig auf einen Verbandplatz geschossen wird - auch mal ein solcher Verbandplatz getroffen wird.

Bei Slawkow kamen wir ins Quartier. Der stark überlegene Feind zieht sich zurück, und wir bleiben auch noch am 22., da es unsere Hauptaufgabe war, die Linie zu halten, in Slawkow. Am 23. blieben wir bis gegen Abend in unserer Stellung und ziehen dann von Slawkow nach Sobota. Auf dem Gut Sobota übernachten wir in einem Hammelstalle, die ganze Batterie, Menschen und Pferde. Die einen von uns haben Andenken von Hammeln, die anderen von Pferden erhalten. - Ueber Sobota ging es am nächsten Tage nach Monkolice. Wieder erhalten wir sehr starkes Infanteriefeuer, verlieren 2 Mann und gehen in Monkolice ins Quartier.

Am 25. geht es fort über Soppel nach Glowno. Immer wieder Gefechte, und als wir am Abend in Glowno ankamen, hofften wir wieder, aber auch wieder vergeblich, etwas längere Ruhe zu bekommen. Schön war unser Quartier in Glowno, - ein Bordell mit zwei mehr oder weniger schmutzigen "Schönen", schreibe ich in mein Tagebuch. Die beiden jungen Damen waren höllisch erstaunt, dass wir so gar keinen Gebrauch von ihnen machten.

Der ursprüngliche Gedanke unserer obersten Heeresleitung war gewesen , geradenwegs von Gostynin auf Lowicz zu marschieren. Da war die Nachricht gekommen, dass unser 25. Reserve-Corps vollkommen vom Feinde eingeschlossen sei, und da war unsere Brigade und eine Kavallerie-Division nach Glowno gesandt worden, um dem 25. Corps Hilfe zu bringen. Dieses unglückliche 25. Corps, das sich im ganzen Kriege so wenig gut geschlagen hat, hatte aber, als wir nach Glowno kamen, sich schon selber herausgehauen und sich nicht nur herausgehauen, sondern eine grosse Zahl von Gefangenen gemacht und viele, viele Geschütze und anderes Kriegsmaterial erbeutet. Unsere Brigade war, wie gesagt, gesandt worden, um dem 25. Corps beizustehen, und die andere Brigade unserer Division hatte den Befehl bekommen, am Abend noch Lowicz zu nehmen. Wenn man bedenkt, dass 4 Corps von uns genau 3 Wochen später Lowicz haben nehmen können, und den Befehl vom 25. vergleicht, so muss man sich sagen: Was werden doch auch manchmal für unausführbare Anforderungen an die Truppen gestellt. Und wenn es jemals eine unausführbare gegeben hat, so war es das Ansetzen unserer 69. Brigade auf Lowicz an diesem Abend. Der Erfolg war natürlich auch gleich Null. Unsere Vorhut ist zwar nach Lowicz hineingekommen, aber auch eben so schnell wieder herausgegangen. Hätte man vielleicht unsere Divisionen damals zusammenlassen können, so wären wir wesentlich schneller vorwärts gekommen, denn in Lowicz haben die Russen nachher alle verfügbaren Kräfte zusam-

mengezogen und erst Lowicz geräumt, als sie einige Kilometer hinter Lowicz bei Skiernewice wieder befestigte Stellungen aufgeworfen hatten. Und so werden sie es auch weiter machen, von Stellung zu Stellung uns kommen lassen, immer wieder sich so befestigen, wie bei Lowicz und nachher bei Skiernewice, um uns das Herankommen nach Warschau möglichst zu erschweren. Denn dichte Wälder in dieser Gegend verhindern ein auch nur einigermassen schnelles Vorwärtskommen, und ungeheure Opfer würde es kosten, - Opfer, zu denen sich unsere Heeresleitung niemals verleiten lassen würde, - wenn man einen solchen Wald im Sturm nehmen würde.

Wir gingen also am 26. zurück über Soppel, Sarry, Sobota wieder nach Zduny in Feuerstellung, aber nur bis zum Abend blieben wir dort, dann ging es wieder auf derselben Strekke auf Glowno, von wo wir am Morgen erst gekommen waren, bis nach Bielawi und dort, in dem netten Städtchen Bielawi verbrachten wir die Nacht. Wie sollte dieses Bielawi 2 Tage später aussehen? ! - Bis 11 Uhr vormittags - wir waren ja erst um 4 Uhr angekommen - hatten wir Ruhe. Plötzlich Alarm; in grosser Eile - dagegen war selbst der Alarm bei Niedzwietzken nichts - raus aus Bielawi. Wir hatten einen Teil unserer Brigade noch in Zduny zurückgelassen und waren ein Infanterie-Regiment und eine Kavallerie-Abteilung stark, als wir aus Bielawi herauszogen und kurz hinter Bielawi in Bielawskiewies in Stellung gingen.

Unsere Abteilung wurde damals von dem Hauptmann der 5. Batterie, Hauptmann Hamann, geführt, der nie viel Liebe für seine Leute gezeigt hatte und dem wir jede artilleristische Fähigkeit absprachen. Wie unfähig er aber auch tatsächlich war, zeigte sich beim Aussuchen der Stellung von Bielawi. Die 5. Batterie musste auf seinen Befehl, obgleich sie eine verhältnismässig gute Stellung hatte, noch weiter vor, und als er von Seiten des Batterieführers Widerspruch hörte, sagte er: "Ich werde Euch schon mal in einen Dreck reinführen, dass Euch Hören und Sehen vergeht." - Der erste, der an diesem Tage fiel, war Hauptmann Hamann. Aber mit seiner Voraussage, dass er uns in den Dreck führen werde, hatte er Recht behalten, denn der schwerste Tag und der verlustreichste für uns, den ich im Feldzuge mitgemacht habe, war dieser Tag von Bielawi, und ist viel von den Verlusten auf sein Konto zu schreiben. -

Um 6 Uhr nachmittags machten die Russen, die in ungefähr dreissigfacher Uebermacht standen, einen Sturmangriff, der wunderbarer Weise zurückgeschlagen wurde. Der Regimentskommandeur und Führer unseres Detachements, Oberstleutnant Müller, ritt nun zum Divisionskommandeur, um Hilfe für uns zu erbitten. Erst für den nächsten Morgen konnte uns Hilfe versprochen werden, und so musste kommen, was gekommen ist: Um 12 Uhr nachts machten die Russen wieder einen Sturmangriff. Zwei Geschütze von uns waren vorn im vordersten Schützengraben -

ich hatte das ja bei anderer Gelegenheit früher beschrieben - unsere Infanterie ging zurück; gehen ist nicht der richtige Ausdruck, - laufen ist nicht der richtige Ausdruck, - sie stürzten zurück. Und unsere beiden Geschütze liessen sie stehen. Wie sie durchgekommen sind und zur Batterie zurück, weiss ich nicht. Es war das Furchtbarste, was man sich aber denken konnte. Zwischen den Russen, deren "Stoy, stoy!" an ihre Ohren klang, hindurch fuhren die braven Kerls. Der einzige, der den Weg wirklich wusste, war Unteroffizier Sengpiel, dem auch die tatsächliche Rettung dieser beiden Geschütze zu danken ist. Wir mussten zurück in eiliger, eiliger Flucht, und nachdem wir 6 bis 7 Kilometer zurückgestürmt waren, hielten wir und gingen wieder von neuem in Stellung. 2 Munitionswagen, 10 Mann und 15 Pferde verloren wir, ein schwerer Tag ! Erst drei Tage später, am 30. November, hatten wir Bielawi wiederbekommen und auch die beiden Munitionswagen noch vorgefunden. - Wie sah aber unsere Infanterie aus ?! Die Leute waren mit ihren Nerven geradezu auf dem Nullpunkt. Und als ich am 30. als Beobachtungsoffizier im vordersten Schützengraben war, war nur immer ein Stöhnen: Na, heute geht es uns sicher wieder so, wie es uns vor drei Tagen gegangen ist. Das ganze Regiment hat 14 Tage später, obgleich wir doch Truppen gewiss nötig brauchten, einen vierzehntägigen Urlaub nach Thorn bekommen. Leutnant Wenz, der den Zug bei Bielawi geführt hatte, bekam verdienter- oder unverdienter Weise (ich lasse das dahingestellt sein,) das Eiserne Kreuz I. Klasse. Unteroffizier Sengpiel hätte es sicher verdient. Aber er bekam später wenigstens das Kreuz II. Klasse dafür. Lange dauert es überhaupt immer, ehe die zum Eisernen Kreuz eingereichten Mannschaften es bekommen. Wenn man bedenkt, dass ich am 7. November eingereicht worden bin und es erst am 10. Dezember erhalten habe,so ist das doch wirklich eine lange Zeit. -

Am 1. Dezember bei Tagesanbruch wurden Horchpatrouillen auf Bielawi gesandt. Jede Sekunde erwarteten wir, dass Schüsse fallen würden, die uns den Beweis bringen sollten, dass der Feind noch Bielawi und die Ortschaften um Bielawi besetzt hielte. Durch das Glas konnte ich den Weg der Patrouillen genau beobachten, aber kein Schuss fiel. Der Feind hatte Bielawi verlassen. Um 10 Uhr vormittags, nachdem die Horchpatrouillen durch Bielawi hindurch waren, zurückkamen und meldeten, dass es wirklich vollkommen frei vom Feinde sei, zogen wir ein und sahen, was aus dem Bielawi in den 3 Tagen geworden war. Es stand noch ein Haus und ein Schornstein, das war alles. Ein grosser, grosser Trümmerhaufen war übrig geblieben. Was hat es uns aber auch gekostet ! - Die verkohlte Leiche eines unserer Leute fanden wir noch. Was aus den anderen geworden, haben wir nicht erfahren, ob sie gefallen oder gefangengenommen sind. Von einen unserer braven Leute, an dem ich sehr gehangen habe, habe ich, Gott sei Dank, später in Berlin gehört, dass er in russische Gefangenschaft geraten sei. Ich will hier meines treuen Kampf- und Schlafgenossen Seebe gedenken, der freiwillig, als am Nachmittag bei Bielawi der Zug schon Verluste gehabt hatte, nach vorn ging, um mitzuhelfen. Hoffentlich bleibt die Anerkennung, die er wahrlich hier verdient hat und über die des öfteren gesprochen worden ist, für ihn nicht aus. -

Bald hinter Bielawi gingen wir wieder in Stellung und beschossen den Feind, den wir noch auf Chroslin abziehen sahen.

Unsere nächste Aufgabe war es nun, Lowicz zu nehmen, was damals, wie ich ja schon erwähnt habe, von einer Brigade genommen werden sollte, und das 4. Corps fast den ganzen Dezember über zu schaffen gemacht hatte. An jenem 1. Dezember wurden wir für eine Nacht zurückgezogen nach Milancwo, um wenigstens etwas Ruhe zu haben. Seit 20 Tagen durften unsere Pferde das erste Mal abgeschirrt werden. - Am 2. Dezember ging es wieder über Bielawi nach Sobota vorwärts. Wohin der Weg führte, wussten wir nicht. Als wir durch Urzyce kamen, erhielten wir auf der Dorfstrasse Infanterie- und Artilleriefeuer und mussten hinter den Häusern Deckung suchen. Das Geschützfeuer hörte aber nicht auf, auch nicht, als eine unserer Batterien bald in Stellung gegangen war, und so entschloss sich unser neuen Abteilungsführer, mein verehrter Hauptmann Krüger, mit der ganzen Abteilung vor Boguria-Gorna in Stellung zu gehen. Beim Aussuchen dieser Stellung, wozu ja immerhin Glück gehört, kam das hervorragende artilleristische Talent unseres Hauptmanns voll zur Geltung. Fast drei Wochen, Tag und Nacht, beschossen uns die Russen, und nicht ein einziges Geschoss kam auch nur in die Nähe unserer Abteilung. Es wurden sogar noch zwei schwere Batterien und eine Mörser-Batterie uns zugeteilt, die die vor Lowicz in Stellung liegenden Russen beschiessen sollten. Meine Zeit in Boguria-Gorna war eine ziemlich eintönige. Es wurde viel geschossen, und manches Dorf, durch das wir vorher durchgezogen waren und durch das wir nachher, als wir von Boguria-Gorna nach Lowicz marschierten, wieder durchkamen, war durch das Geschützfeuer unserer Batterie zerstört worden. Jede dritte Nacht musste ich mit einem Zuge nach vorn, nach Urzyce, um, falls ein feindlicher Angriff erfolgen sollte, unsere Infanterie zu unterstützen. Aber in der ganzen Zeit hat kein Angriff stattgefunden. Am 4. Dezember versuchen wir erfolglos, einen feindlichen Flieger herunterzuschiessen, der auch unsere Artilleriestellung verrät und wir deswegen mit schwerer Artillerie beschossen werden. Aber niemals ist eins von den Geschossen zu dicht an uns herangekommen. Bald lagen sie weit vor uns, bald hinter uns, aber, Gott sei Dank, nie an der richtigen Stelle.

Hier in Boguria-Gorna wurden wir auch noch gegen Cholera geimpft. Eine Impfung auf Typhus habe ich dort nicht mehr mitgemacht.

Am 10. Dezember nachmittags, als wir aus der Feuerstellung kamen, wurde ich zum Hauptmann gerufen, der mir sagte, dass ich mich abends um 6 Uhr im Quartier des Majors, der inzwischen wieder zur Abteilung zurückgekehrt war, zu melden hatte. Da ich mir keiner Schuld bewusst war (und wenn ich etwas ausgefressen hätte, mir es der Hauptmann wohl sicherlich gesagt hätte,) nahm ich an, dass es sich sicherlich um die Belohnung für Suwalki handeln würde. Und ich hatte mit meiner Vermutung nicht unrecht. Als ich zum Major kam, überreichte er mir und noch drei Anderen mit einer längeren Ansprache das Eiserne Kreuz, und als er uns nachher gratulierte, sagte er zu mir: "Was wird jetzt wohl Frau Riess sagen ?" - Ich hatte ja zu Anfang schon erwähnt, dass mein Major früher in meiner Einjährigenzeit mein Batterieführer gewesen war, der meine Frau und mich in meiner Einjährigenzeit auf einer Urlaubsreise in Kissingen einmal getroffen hatte. Und die Art, wie er mir das Kreuz überreichte, war eine wirklich entzückende.

Am 13. schossen die Russen ein Gehöft in Boguria-Gorna in Brand und wir glauben nun, dass sie unser Quartier genau ausfindig gemacht haben. Aber als am Abend kein weiterer Schuss fiel, bezogen wir wieder unser altes Quartier. Man wird doch der Gefahr gegenüber unglaublich dickfellig, wenn man bedenkt: die Russen wissen, wo man liegt, und man legt sich ruhig wieder schlafen. -

Eine österreichische Motor-Batterie war auf unseren linken Flügel gekommen und sandte den Lowiczer Bewohnern ihre Grüsse hinüber. Am 14. machen die Russen einen starken Angriff auf unseren rechten Flügel, sodass wir alles zum Abmarsch fertigmachen müssen. Aber auf unserer im Zentrum gelegenen Stellung bleibt alles ruhig. Am Abend können wir den unglaublich schweren Kampf auf dem rechten Flügel beobachten. Bei diesem Sturm haben die Russen das Dorf und den Kirchhof Chroslin genommen, und am 15. wird von unserem rechten Flügel ein Gegenangriff gemacht, der von unserer Artillerie unterstützt wird. Dorf sowohl wie Kirchhof von Chroslin haben wir wieder zurückgewonnen. Dieser Sturmangriff vom 14. und 15. hatte aber einen ganz besonderen Zweck, der für uns Deutsche einfach unverständlich ist. Sowohl am 14., wie am 15. hatten die Russen ungeheure Verluste. Regimenter über Regimenter wurden vor unsere Maschinengewehre gehetzt. Und der Zweck ? - Die Russen wollten ungestört aus Lowicz abziehen, ihr Kriegsmaterial in Sicherheit bringen, und da durften Menschen keine Rolle spielen. Sie haben natürlich ihren Zweck erreicht, denn wer konnte annehmen, dass, während Sturmangriffe über Sturmangriffe gemacht werden, die Russen abziehen ? - Es war dadurch auch auf dem linken Flügel und bei den Truppen, die Lowicz gerade gegenüber lagen, garnicht der Versuch zum Vorgehen gemacht worden. Am 16. schrieb ich in mein Tagebuch: Es ist alles ruhig, fast als wenn die Russen abziehen wollen. - Es wurden deswegen Horchpatrouillen hinausgeschickt, die mit ganz verschiedenen Meldungen zurückkamen. Die einen meldeten, dass die Russen alles zum Sturmangriff im Zentrum fertigmachten. Daraufhin macht sich unsere Infanterie bereit, um dem Sturm entgegentreten zu können. Die zweite Horchpatrouille meldet: Es ist alles ruhig. Und die dritte meldet: Man hört fortlaufendes Wagengerassel. Kein Schuss fällt, was in den ganzen drei Wochen noch nicht vorgekommen war. Und als nun auch am 17. alles ruhig bleibt, werden am Tage neue Patrouillen vorgeschickt, die nicht beschossen werden, und so ziehen wir , nachdem wir 17 Tage in Boguria-Gorna in dem Hause der Polenfürstin Matka, wie wir die Besitzerin der elenden Stinkbude nannten, in der wir hausten, fort. Am Abend noch trafen wir in Lowicz ein. Die Truppen, die vor uns dort angekommen waren, hatten noch mit der Nachhut des Feindes einen Kampf. Als wir einzogen, war aber schon alles ruhig. In Lowicz hatten wir ein gutes Quartier und konnten die Nacht ruhig schlafen. Sehen dort die unglaubliche Wirkung der österreichischen Motorbatterien, die Löcher auf den Marktplatz und die Strassen von Lowicz gerissen haben von einer Breite von 8 Metern und einer Tiefe von 3 Metern.

Am 18. gehen wir bei Sirakowice in Stellung. Natürlich haben wir wieder mit dem Feinde zu tun, der den Rückzug der Russen decken soll, die sich hinter Skiernewice in ganz feste Stellungen begeben haben. Am 19. ziehen wir durch Skiernewice, wo sich ein wunderbares Jagdschloss des Zaren befindet. Die Stadt ist im Gegensatz zu den meisten anderen russischen Städten verhältnismässig sauber. - Hinter Skiernewice kommen wir sofort ins Gefecht und bleiben in der eisigen Kälte die ganze Nacht draussen. Am nächsten Tage bleiben wir in unserer Stellung und beziehen in Schwabeln Quartier. Dies ist eine kleine Kolonie, ganz in der Nähe von Skiernewice, die von deutschen Einwanderern gegründet wurde. Auch heute noch sprechen die Nachkommen dieser Deutschen noch etwas deutsch, dass man sich wenigstens mit einigen von ihnen verständigen konnte.

Diese Stellung der Russen, die sie hinter Skiernewice eingenommen hatten, zu nehmen, war einfach unmöglich, denn Schützengraben lag hinter Schützengraben und Drahtverhau hinter Drahtverhau. Wie weit wir heute an dieser Stelle sind, weiss ich nicht. Wir rückten am 21. noch 2 Kilometer gegen den Wald vor, erhalten beim Einfahren starkes Artilleriefeuer, und wenn die Russen mehr Geschütze dort gehabt hätten, wäre es uns sicherlich schlecht gegangen. Wir blieben in unseren guten Quartieren in Schwabeln, während wir am Tage draussen in derselben Stellung blieben, denn ein Vorwärtsgehen über den Fluss und in den Wald hinein ist ein Ding der Unmöglichkeit. Schwere Geschütze von uns sind auch in dieser Stellung aufgestellt worden, und interessant war es mir, zu beobachten, wie unsere 21-cm-Geschütze, denn solche waren es, bedient wurden. Es ist *eine unglaubliche Detonation, wenn solch ein Ding losgeht. Und die Handhabung dieser Geschütze ist trotz der Grösse eine so einfache, dass man es wirklich immer wieder und wieder bewundern muss, was die Technik hervorbringt.*

Unsere schweren Brummer schlagen in die feindlichen Schützengräben und der Erfolg ist der, dass ein feindliches Bataillon sich ergeben will und mit erhobenen Händen auf uns zu kommt. Da werden die eigenen Maschinengewehre der Russen gegen diese Mannschaften angesetzt und die Leute von ihren eigenen Stellungen aus niedergemäht. Wenn das der Geist ist, der die Russen beseelt, dann werden sie es wohl nicht mehr lange aushalten. Aber was ist eben Menschenmaterial bei den Russen ? ! - Wenn heute die Leute durch Anstrengungen erschöpft sind, werden sie zurückgezogen und durch neue Truppen ersetzt - Menschen spielen keine Rolle. -

An der Weihnachtsfeier, die draussen im Walde stattfand, nahe der Stellung, kann ich schon nicht mehr teilnehmen, da ich mir eine böse Erkältung zugezogen habe, und ich blieb in dem Hause in Schwabeln bis zum 29. Dezember. Am 29. nachmittags um 5 Uhr ziehen wir auf den linken Flügel unserer Stellung, wo den ganzen Tag schon schwer gekämpft wurde,und kommen nachts um ½ 2 Uhr nach Borowino ins Quartier. An diesem Tage fühlte ich mich schon sehr, sehr elend. Und als ich am nächsten Morgen losziehen wollte, kam auf Wunsch eines meiner Ka-

meraden der Arzt, der mir dringend empfahl,mich krank zu melden, um mich nicht vollkommen zu ruinieren. Aber ich konnte mich selbstverständlich dazu noch nicht entschliessen, obgleich ich sicher hohes Fieber hatte. Am 30. noch hielt ich es aus. Wir sollten in der Nähe von Bolimow in Stellung gehen, werden aber hinter einem kleinen Dorf als Reserve aufgestellt. Dorthin kommt unser kommandierender General, General von Morgen, und begrüsst dort unsere 36. Reserve-Division, lässt sich jeden einzelnen, mit dem Eisernen Kreuz dekorierten Mann vorstellen und feuert alle an, weiter ihre Pflicht zu tun, bis der Feind vollkommen niedergerungen sei. Dann reiten wir nach Pyarski ins Quartier.

Am nächsten Morgen, als ich aufwachte, konnte ich nicht einen Ton mehr sprechen und musste nun wohl oder übel zum Hauptmann gehen, der mir einen Urlaub anbot, damit, wie er sagte, ich der Batterie nicht verloren gehen sollte. Ich war der erste in der Abteilung, der überhaupt Urlaub bekommen hat, denn der Major genehmigte ohne weiteres die Bitte unseres Hauptmanns, und so ritt ich denn, begleitet von meinem treuen Sell und meinem lieben alten Fuchs, den ich den grössten Teil des Feldzuges geritten habe, nach Lowicz. Und als ich in Lowicz nach dreistündigem scharfen Ritt ankam, ging gerade eine Stunde später ein Verwundetentransport ab, der mich zu einem Urlaub nach Deutschland bringen sollte. Durch die Sylvesternacht fuhr ich nach Thorn. Ich war zu kaput, um jetzt schon das hohe Glücksgefühl des Nachhausekommens zu haben, aber als ich wieder über die deutsche Grenze bald hinter Wloclawek kam, da hatte ich doch das wunderbare Gefühl: jetzt kommst du heim !- Aber tagelang nachher noch, nachdem ich schon zuhause war, hörte ich immer schiessen, und wenn eine Kleinigkeit zu Boden fiel oder wenn irgendwie laut gesprochen wurde, so schreckte ich zusammen, was ich doch draussen nie gekannt habe, - so waren doch meine Nerven ermattet.

Am 1. Januar vormittags 11 Uhr traf ich in Thorn ein. Unterwegs begegnete ich noch ungeheuren Transporten, die raus an die Front kamen. Und bei allen eine solche Siegeszuversicht, dass man wirklich denken kann: wo Leute mit solcher Begeisterung hinausziehen, da kann der Erfolg unmöglich ausbleiben.

Von Thorn aus fuhr ich am Mittag nach Berlin, wo ich am Abend eintraf. Was soll ich über meine Ankunft in Berlin sagen ? - Was ich auch sagen würde von der Glückseligkeit, die ich empfand, als ich zu Hause war, es wäre zu wenig. Ich war eben zuhause ! Nach fünf Monaten schwerer Kämpfe, anstrengender Märsche und schwerer, schwerer Entbehrungen. Wie viel von denen, mit denen ich hinausgezogen bin, mit denen ich Tag und Nacht zusammengewesen bin, sind noch übrig geblieben ? - Wenige, wenige nur, denn den Kämpfen, die wir zusammen mitgemacht hatten, folgten noch unsagbar schwere. Und wer nicht tot, verwundet oder gefangengenommen ist, schied, so wie ich, durch

Krankheit aus.Denn als mein Urlaub beinahe vorüber war und ich mich wieder zu meiner Batterie zurückbegeben wollte, setzte mein altes Leiden, mit dem ich ja schon vor dem Kriege zu tun hatte, mit grosser Heftigkeit wieder ein und macht es mir unmöglich, weiter an dem grossen Ringen, wenigstens vorläufig, teilzunehmen.

Wenn ich nun alles, was ich in diesen meinen Kriegserlebnissen gesagt habe, zusammenfasse, so sieht man am besten daraus, ein wie glänzender Geist in unseren Truppen steckt, wie alle Strapazen und Anstrengungen bei einem Vorwärtsgehen leicht überwunden werden, und wenn aus diesem oder jenem Grunde es nötig ist, dass unsere Truppen zurückgenommen werden, uns alle nur der eine Wunsch beseelt hat: Wieder vorwärts !

Ein Krieg, der mit solchen Truppen und unter solchen Führern geführt wird, der muss zu dem grossen, wunderbaren Erfolg führen, der uns allen vor Augen steht. Und wenn man auch heute noch auf den Bahnhöfen sieht, mit welcher Begeisterung unsere jungen Truppen hinausziehen, dann kann man mit vollem Vertrauen sagen:

Lieb Vaterland, magst ruhig sein !

Below: from Gertrud Riess's diary, in her handwriting, the three stanzas that Werner sent to her before Christmas, 1914. On the following two pages are the stanzas she wrote and sent back to him. They all are alternative lyrics for the Austrian Christmas carol, *Stille Nacht (Silent Night).*

„O stille Nacht, o heilige Nacht
Wo höher schlagen alle Herzen,
Wo heißer brennt der Sehnsucht Schmerzen
Und einer, einer nur hab ich gedacht! –

So fern ist sie an Osten mir,
Und in mir lebt doch nur ihr Denken!
Wann, Herr, wirst Du uns Frieden schenken?
Wann wird sie wieder sein bei mir?

O stille Nacht, o heilige Nacht,
Die Du den Frieden bringst auf Erden,
Wird Frieden, goldner Frieden werden?
– Gib ihn doch, Gott, in Deiner Macht! –

Am selben Abend, an dem ich dies kleine Weihnachtsgedicht erhielt, schrieb ich ihm folgende Zeilen:

„Stille Nacht, heilige Nacht,
Alles schläft, einsam wacht
Nur das traute, hochheilige Paar,
Holder Knabe im lockigen Haar,
Schlaf in seliger Ruh, schlaf in seliger Ruh!"

Sangen so Kinderlein froh,
Süß erklang ihr Gesang,
Wenn beim strahlenden Kerzenschein
Freude zog in die Herzen hinein
In der heiligen Nacht, in der heiligen Nacht.

Stille Nacht, heilige Nacht,
Sehnsuchtsvoll Liebster wacht,
Denkt voll Wehmut nach Haus, nach Haus,
Steht allein jetzt in Kampfesgebraus
In der heiligen Nacht, in der heiligen Nacht.

Stille Nacht, heilige Nacht,
Herzensbang Liebste wacht,
Denkt in bedrückender Einsamkeit,
Wehmutsvoll der vergangnen Zeit
In der heiligen Nacht, in der heiligen Nacht.

Stille Nacht, heilige Nacht,
Siehe da, sacht, ganz sacht,
Zukunftstraum sich dem Auge enthüllt:
Geh im Frieden, – glückeserfüllt
Uns in Liebe vereint, uns in Liebe vereint! —

ABOUT THE AUTHOR

Werner Riess was born in Stettin (now Szczecin, Poland) on November 21, 1887. He completed his mandatory three years of military service in the German field artillery, then moved to Berlin, worked as a manager in a large department store, and married Gertrud Grumach, who lived three doors down the street. In the summer of 1914 they lived in Schöneberg. He was twenty-six years old and had become a partner in the department store.

However, Riess was a corporal in the army active reserves and was sent to the Eastern Front just before the Russians invaded in August 1914. He fought with a newly created field artillery battery, was promoted to cavalry sergeant, rewarded the Iron Cross for bravery, but became so ill in late December that he was sent to central Germany for surgery. Recovering in the hospital, he wrote his memoir while it was still fresh in his mind and reported to headquarters in Berlin where he continued his service. In 1916 he and Gertrud had a son.

Werner Riess died in 1918.

ABOUT THE EDITOR

Warren Riess is Research Associate Professor of History, Emeritus at the University of Maine. He enjoyed a forty-five year career as a maritime historian and archaeologist and now lives on a Maine coastal peninsula with his wife. He is the grandson of Werner Riess and was named after him. His website is warrenriess.com.

www.ingramcontent.com/pod-product-compliance
Lightning Source LLC
Chambersburg PA
CBHW031126090426
42738CB00008B/989

* 9 7 8 0 9 7 1 3 4 3 8 1 8 *